CELL-FREE PROTEIN SYNTHESIS

Edited by **Manish Biyani**

Cell-Free Protein Synthesis
http://dx.doi.org/10.5772/2955
Edited by Manish Biyani

Contributors

Maximiliano Juri Ayub, Walter J. Lapadula, Johan Hoebeke, Cristian R. Smulski, Greco Hernández, Manish Biyani, Madhu Biyani, Naoto Nemoto, Yuzuru Husimi, Kodai Machida, Mamiko Masutan, Hiroaki Imataka, Takanori Ichiki, Tokumasa Nakamoto, Ferenc J. Kezdy, Assaf Katz, Omar Orellana

Published by InTech

Janeza Trdine 9, 51000 Rijeka, Croatia

Publishing Process Manager Marijan Polic
Technical Editor InTech Prepress, Novi Sad
Cover InTech Design team

Additional hard copies can be obtained from orders@intechopen.com

Cell-Free Protein Synthesis, Edited by Manish Biyani
p. cm.
ISBN 978-953-51-0803-0

For EU product safety concerns:
IN TECH d.o.o., Prolaz Marije Krucifikse Kozulić 3, 51000 Rijeka, Croatia, info@intechopen.com or visit our website at intechopen.com.

Meet the editor

Manish Biyani was educated at Saitama University (Ph.D. in Biophysics) and The University of Tokyo (Postdoctoral in Nanobioengineering) in Japan. He has pursued a career in basic research in biotechnology and its extended application in various fields including medicine, healthcare. He is currently Project Professor at Department of Bioengineering at The University of Tokyo in Japan and is Founding Research Director for the Biyani Group of Colleges in India. He is also involved to promote scientific interaction and research networks between India and Japan. He is the recipient of numerous prestigious awards including JB OUP award from International Union of Biochemistry and Molecular Biology, Bursary award from International Genetics Federation, MEXT scholarship award from Japanese Ministry of Science & Technology, Outstanding academic award from Pharmacy Medical Association of Rajasthan. He is also an ad hoc reviewer in many scientific journals.

Contents

Preface

A half century ago, Nierenberg and Matthaei discovered the first codon UUU for phenyl alanine using a cell-free translation system from DNase treated *E.coli* extract. From that time on, the cell-free protein synthesis has been used for the analysis in molecular biology, protein production and protein design, taking advantage of compartment-free experiment. Post-genome proteomics and functional genomics require a high throughput systematic production of proteins. Evolutionary protein engineering requires the translation of a large diversity library. Studies on the translation itself (its molecular mechanism, its origin etc.) require a simplified model system. Cell-free protein synthesis systems are useful for all these themes. This book reviews briefly the history of the translation and the history of its study.

One of the most astonishing molecular events which molecular biology has discovered is the translation process, that is, a Natural digital-to-digital decoding process. Structural biology found the ribozymatic peptidyl transferase action of ribosome and finally gave us the concept of RNA-makes-Protein. And it suggested the RNA+Protein world was emerged from the RNA world. In the RNA world, the *molecular coding* process was established probably due to three folds complementarities of RNA molecules as follows: (i) complementary base pairing for amplification, (ii) complementary base pairing for folding and (iii) the complementarity between the surface of the folded RNA and a ligand molecule. The first is related to genotype and the second plus the third are related to phenotype. The phenotype as the molecular function (e.g. specific binding to the ligand) could be digitally encoded in the genotype as the base sequence of the RNA molecule, through the Darwinian selection process just as that of exploiting an RNA aptamer. On the other hand, the decoding process is very simple, i.e., folding and binding. This is the digital-to-real-world decoding. The above mentioned encoding process is also not so complicated, due to the RNA-type genotype-phenotype linking strategy, that is, both on the same molecule. Evolvability of RNA is based on this molecular coding ability. Using this evolvability, evolutionary RNA engineering in these two decades have been creating many kinds of functional artificial RNAs (including new drugs!) and thus indicated the potentiality of RNA molecules and the physico-chemical possibility of the RNA world.

Linking with nucleic acids, polypeptide finally got evolvability and was able to become proteins. The genotype-phenotype linking strategy for Darwinian selection of protein is not so simple. There are three types of the strategy in evolutionary protein engineering as follows: the virus-type, the cell-type and the external intelligence-type. In the virus–type, mRNA and its protein are bound together just as in the simplest virus particle. In the cell-type, mRNA and its protein are in a same compartment, *e.g.* a bacterial cell or a micro plate well. In the origin of the translation, what strategy was adopted is not clear. The most complicated aspect of the translation, however, may be the digital-to-digital *decoding* process. Note: there is no problem in the digital-to-digital *encoding* because there is no reverse-translation. The encoding process is accomplished via the Darwinian selection using this digital-to-digital decoding and the above mentioned genotype- phenotype linking.

The central issue in the origin of the translation is the establishment of the genetic code table for the digital-to-digital decoding. Present-day standard genetic code table seems to be evolutionally optimized if we admit our twenty amino acids. But whether twenty and these twenty were optimal or not is open question. In fact, protein engineers have been introducing many kinds of non-natural amino acids, tricking the code table for their purposes. What is the primitive ribosome is also open question. But there is a primitive tRNA model as the first gene. Anyway there should have been a co-evolution process of RNA replication and the primitive translation. There are evidences to suggest common unit processes in both RNA replication and the translation. Present-day standard genetic code table is almost universal on the Earth. And the translational apparatuses are the most conservative molecular machines. These indicate the bottleneck of the biological evolution on the Earth was the establishment of the translation. The enhancement of evolvability of an organism by introducing evolving proteins must overbalance the difficulties of passing the bottleneck. Thus, protein biosynthesis had two important aspects from the beginning as a matter of course: innovative molecular design and regulated production. These two aspects are also important for modern protein engineers.

The editor of this monograph, Dr. Manish Biyani, is an innovative researcher in the field of cell-free protein synthesis, evolutionary protein engineering and experimental genome analysis. I hope readers enjoy the scope of Dr. Biyani and splendid informative chapters by expert scientists contributed in this book.

Preface written by:
Yuzuru Husimi
Prof, Saitama University,
Japan

Preface confirmed by:
Manish Biyani
Prof, The University of Tokyo,
Japan

Section 1

Fundamental Understanding and Protein Synthesis

Chapter 1

Ribosomes from Trypanosomatids: Unique Structural and Functional Properties

Maximiliano Juri Ayub, Walter J. Lapadula, Johan Hoebeke and Cristian R. Smulski

Additional information is available at the end of the chapter

http://dx.doi.org/10.5772/48336

1. Introduction

Trypanosomatids are a monophyletic group of protozoa that diverged early from the eukaryotic lineage, constituting valuable model organisms for studying variability in different highly conserved processes including protein synthesis. Moreover, several species of trypanosomatids are causing agents of endemic diseases in the third world. There are many evidences suggesting that translation in these organisms shows important differences with that of model organisms such as yeast and mammals. These unique features, which have a great potential relevance for both basic and applied research, will be discussed in this chapter.

2. Structural analysis

2.1. Cryo-electron microscopy map of *Trypanosoma cruzi* ribosome: Unique features of the rRNA

Using the cryo-electron microscopy (cryo-EM) technique, a 12Å resolution density map of the *T. cruzi* 80S ribosome has been constructed [1]. The overall structure of the *T. cruzi* 80S ribosome exhibits well defined small (40S) and large (60S) subunits (Figure 1). Some of the landmark characteristics of the ribosome structure can be identified in the density map. Compared with the 80S ribosome from yeast, both the small and large ribosomal subunits from *T. cruzi* are larger, mainly due to the size of the ribosomal RNA molecules. *T. cruzi* rRNA (18S rRNA: 2,315 nt and 28S rRNA: 4,151 nt) is one-fifth larger than yeast rRNA (18S rRNA: 1,798 nt; 25S rRNA: 3,392 nt) in total number of nucleotides.

Although the *T. cruzi* 80S ribosome possesses conserved ribosomal structures, it exhibits many distinctive structural features in both the small and large subunits. Compared with other eukaryotic ribosomes, the *T. cruzi* ribosomal 40S subunit appears expanded, due to the addition of a large piece of density adjacent to the platform region (Figure 1). As can be seen in the secondary structure of the *T. cruzi* 18S rRNA (Figure 2), this extra density must be attributed to two large expansion segments (ES) in domain II of the 18S rRNA, ES6 and ES7, designated as insertions of helices 21 and 26. These are the two largest ES in the *T. cruzi* 18S rRNA, involving 504 and 147 nucleotides, respectively.

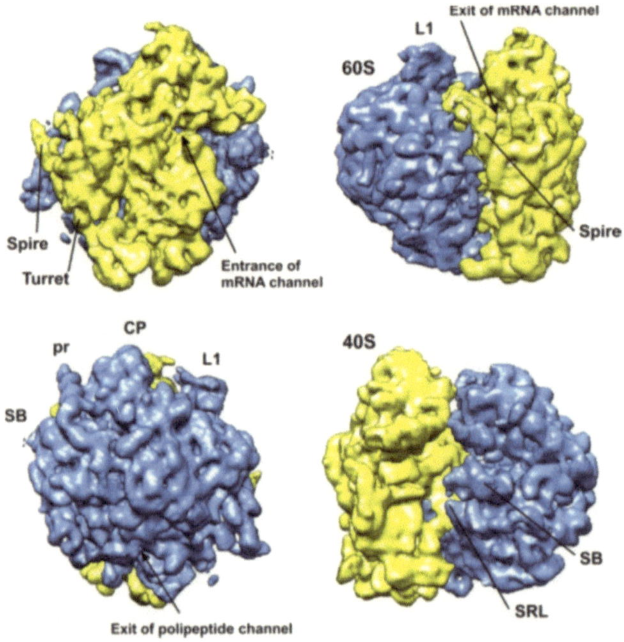

Figure 1. Cryo-electron microscopy map of *T. cruzi* 80S ribosome. Blue: large subunit. Yellow: small subunit. Landmark characteristics are indicated: SB, stalk base; SRL, sarcin-ricin loop; L1, L1 protein; CP, central protuberance; pr, prong.

Part of ES6/ES7 makes up a large helical structure (named the "turret"), located at the most lateral side of the 40S subunit (Figures 1 and 2). The turret measures 205 Å in length and forms the longest helical structure ever observed in a ribosome. The upper end of the turret appears as a sharp, freestanding spiral of 50 Å in length, named "spire," located next to the exit of the mRNA channel. The distance between the spire and the mRNA exit is ~130 Å. The lower portion of the turret extends all of the way to the bottom of the 40S subunit. At its lower end, it bends by almost 90° and forms a bridge with the 60S subunit. This is a unique type of connection between the small and large subunits, as compared with all other ribosomal structures investigated to date [2]. Apart from the turret, the extra density in the 40S subunit also includes several small helical structures as part of ES6 and ES7. These helical structures observed in the density map are in accordance with the comparative analysis result based on ES6 sequences from >3,000 eukaryotes, in which several helices were identified only in kinetoplastida [3]. The ES3, ES9, and ES10 are located near helices 9, 39, and 41, respectively, and are associated with three small masses in the density map of the 40S ribosomal subunit, one at the bottom of the 40S ribosomal subunit, the other two in the head region.

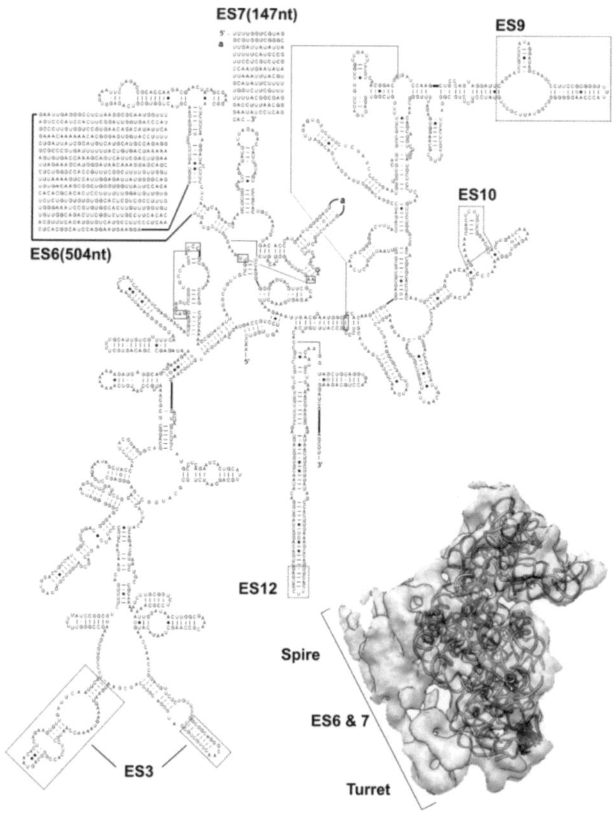

Figure 2. Secondary structure of *T. cruzi* 18S rRNA with the characteristic ES. The 40S subunit was superposed with the crystallized *S. cerevisiae* 18S rRNA (PDB: 3U5E). The volume occupied by ES6 and ES7 is indicated.

In contrast to the other rRNA regions, ES12, located in the long penultimate helix from 18S rRNA, is shorter in *T. cruzi* compared to other eukaryotes. This results in the helix 44 in *T. cruzi* (113 nt) being longer than in *E. coli* (103 nt), but shorter than in yeast (129 nt). Consistently with these variations in length, the span of the density attributable to helix 44 in the *T. cruzi* ribosome also has an intermediate position between *E. coli* and yeast. Interestingly, this region forms the decoding center, and is also the action site of aminoglycoside antibiotics. Moreover, differences in this region have shown to be responsible for the higher susceptibility of trypanosomatid ribosomes to the aminoglycoside paromomycin [4], as will be discussed below.

The *T. cruzi* 60S subunit contains several large extra densities located at its periphery, in contrast with the large ribosomal subunit in yeast. The most common structures can be identified, such as the central protuberance (CP), L1 stalk, and stalk base of P proteins, as well as the conserved rRNA core structure (Figure 1). Although the secondary structure of the *T. cruzi* rRNAs in the 60S subunit is not available, the locations of most of the observed extra densities are consistent with the general locations of the rRNA expansion segments, which are, as a rule, at the surface of eukaryotic ribosomes. Among the extra densities, there is a large helical structure (''prong'') located between the CP and helix 38, in the back of the

60S subunit (Figure 1, pr). Interestingly, a similar feature was reported only in the structure of the human ribosome, but not in yeast neither bacterial ribosomes [5]. Morphological comparison of the 60S ribosomal subunit from *T. cruzi* with those from yeast and higher eukaryotes reveals that the *T. cruzi* 60S ribosomal subunit does not possess the universal eukaryotic feature of a planar surface near the exit site of polypeptide [2]. Instead, the 60S ribosomal subunit from *T. cruzi* presents a shape that is similar to those from bacteria. In contrast to the conserved eukaryotic rRNA core structure, the location of the L1 stalk in *T. cruzi*, which is on one side of the CP, does not match either of the two reported positions in yeast, known as the "in position" and "out position," in relation to the ratchet-like subunit rearrangement. Instead, the L1 stalk in *T. cruzi* takes an in-between position, possibly due to its high mobility. On the other side of the CP, the P protein stalk is not visible in the density map of the *T. cruzi* 60S ribosomal subunit, whereas Western blots of the ribosome preparation using monoclonal antibodies against P proteins (P0/P1/P2) showed that these proteins were present in the ribosome preparation. As homologs of the bacterial moiety $L10/(L7/L12)_4$, P proteins are known to be very flexible, and the absence of a stalk in the cryo-EM density map is likely due to the lack of stabilization. A complete description of *T. cruzi* stalk region, components, interactions and complex formation will be discussed below.

2.2. Sequence and proteomic analysis: Differences on the ribosomal proteins

The Cryo EM map of *T. cruzi* ribosomes exposed important differences in comparison with the corresponding organelles of model organisms such as *S. cerevisiae* and mammals [1]. Some of them were attributed to large expansions in the primary sequence of the ribosomal RNA molecules. However, the presence of specific features due to ribosomal proteins is difficult to demonstrate by this technique. Therefore, using the *S. cerevisiae* ribosomal protein sequences as probes, it was possible to identify in the *T. cruzi* genome database all homologue genes [6]. The average amino acid identity between the *S. cerevisiae* and *T. cruzi* ribosomal proteins was remarkably low (~50%), taking into account the high degree of conservation of the ribosome through evolution.

The ribosomal proteins inferred by data mining were compared to the MS analysis results from whole parasites [7] and purified ribosomes [6]. Results are summarized in Tables 1 and 2 for proteins from the large and small subunits, respectively.

S. cerevisiae			*T. cruzi*			
Prot	3U5E	Length (aa)	Length (aa)	% ID	Prot MS	Ribo MS
L1	-	217	214	51	+	-
L2	+	254	260	62	+	+
L3	+	387	428	57	+	+
L4	+	362	374	49	+	+
L5	+	297	309	49	+	+
L6	+	176	193	43	+	+
L7	+	244	242	41	+	+
L8	+	256	315	47	+	+

S. cerevisiae			*T. cruzi*			
Prot	3U5E	Length (aa)	Length (aa)	% ID	Prot MS	Ribo MS
L9	+	191	189	46	+	+
L10	+	221	213	61	+	+
L11	+	174	192	69	+	+
L12	-	165	164	56	+	+
L13	+	199	218	39	+	+
L14	+	138	180	30	+	+
L15	+	204	204	57	+	+
L16	+	199	222	44	+	+
L17	+	184	166	54	+	+
L18	+	186	193	43	+	+
L19	+	189	357	50	+	+
L20	+	174	179	37	+	+
L21	+	160	159	41	+	+
L22	+	121	130	33	-	+
L23	+	137	139	69	+	+
L24	+	155	125	32	+	+
L25	+	142	226	45	+	+
L26	+	127	143	57	+	+
L27	+	136	133	43	+	+
L28	+	149	145	59	+	+
L29	+	59	71	82	+	+
L30	+	105	105	55	+	+
L31	+	113	188	42	+	+
L32	+	130	133	42	+	+
L33	+	107	149	42	+	+
L34	+	121	170	38	+	+
L35	+	120	127	44	+	-
L36	+	100	114	41	+	+
L37	+	88	84	57	+	+
L38	+	78	82	43	-	+
L39	+	51	51	60	-	-
L40	+	52	52	65.4	+	-
L41	+	25	Not Found			
L42	+	106	106	65	-	+
L43	+	92	90	61	+	+

Table 1. Proteins from the large subunit. Left, *S. cerevisiae*: protein name, presence in the crystal structure (PDB: 3U5E) and number of residues. Right, *T. cruzi* homologues: amino acid length, percentage of identity and positive (+) or negative (-) detection by MS on whole parasites (Prot MS) or purified ribosomes (Ribo MS).

S. cerevisiae			*T. cruzi*			
Prot	3U5C	Length (aa)	Length (aa)	% ID	Prot MS	Ribo MS
S0	+	252	245	52	+	+
S1	+	255	261	40	+	+
S2	+	254	263	58	+	+
S3	+	240	214	58	+	+
S4	+	261	273	50	+	+
S5	+	225	190	64	+	+
S6	+	236	250	49	+	+
S7	+	190	211	34	+	+
S8	+	200	221	47	+	+
S9	+	197	190	59	+	+
S10	+	105	161	34	+	+
S11	+	156	173	54	+	+
S12	+	143	142	30.8	+	+
			141	34.5	+	-
S13	+	151	151	62	-	+
S14	+	137	144	74	+	+
S15	+	142	152	53	+	+
S16	+	143	149	56	+	+
S17	+	136	141	57	+	+
S18	+	146	153	58	+	+
S19	+	144	167	35	+	+
S20	+	121	117	41	+	+
S21	+	87	251	43	+	+
S22	+	130	130	72	+	-
S23	+	145	143	68	+	+
S24	+	135	137	47	+	+
S25	+	108	110	39	+	-
S26	+	119	112	42	+	+
S27	+	82	86	61	+	+
S28	+	67	91	68	-	+
S29	+	56	57	54	+	+
S30	+	63	65	65	-	-
S31	+	152	150	60	+	+
RACK1	+	319	317	43	+	+

Table 2. Proteins from the small subunit. Left, *S. cerevisiae*: protein name, presence in the crystal structure (PDB: 3U5C) and number of residues. Right, *T. cruzi* homologues: amino acid length, percentage of identity and positive (+) or negative (-) detection by MS on whole parasites (Prot MS) or purified ribosomes (Ribo MS).

This analysis showed that *T. cruzi* ribosomal proteins are, in average, longer than the corresponding *S. cerevisiae* proteins. The extra regions in *T. cruzi* ribosomal proteins are generally at the N- or C-terminal ends. The most intriguing examples of these terminal extensions, when comparing to yeast, are TcL19 and TcS21 (blue rows on Tables 1 and 2, respectively), showing C-terminal extensions of 168 and 164 amino acids, respectively. These extensions are only present in kinetoplastids, although their length varies among species. MS analyses of *T. cruzi* ribosomes confirmed the presence of peptides matching to TcL19 and TcS21, strongly suggesting that these genes correspond to the functional ribosomal components [6]. The possible functional roles of these extensions, as well as the molecular mechanisms that generated them over time, constitute interesting fields for future studies.

It is interesting to note that *S. cerevisiae* L19 protein has been described as forming part of the polypeptide chain exit channel [8]. In addition to L19, the polypeptide chain exit channel is formed by L17, L25, L26, L31 and L35. All of these proteins show important extensions in *T. cruzi*, ranging from 41 amino acids (L35) up to 57 amino acids (L26). This fact can be related to the absence of a flat surface on this region in *T. cruzi* 80S ribosome, in contrast to the corresponding region in the yeast ribosome [1].

Two putative homologue genes for the S12 protein sharing 65% of amino acids identity are present in the *T. cruzi* genome (named TcS12A and TcS12B). TcS12A is slightly closer to yeast S12 (34.5% of amino acids identity) than TcS12B (30.8% of amino acids identity). Both genes were expressed at the protein level [7] but only TcS12B was detected in the proteomic analysis of purified ribosomes (Table 2). Interestingly, there are also two genes in *T. brucei* (S12A and S12B) but only one in *L. major*, suggesting a gene duplication event after the divergence of *Leishmania* spp into the trypanosomatid lineage.

In other eukaryotes, such as mammals and yeast, ribosomal proteins S31 and L40 are synthesized as a C-terminal fusion with ubiquitin. Data mining revealed similar fusion genes in the *T. cruzi* genome. From these ubiquitin fusion proteins, only TcS31 (also named S27A) was detected by mass spectrometry on pure ribosomes.

Out of the 32 proteins found by sequence identity to *S. cerevisiae* 40S proteins, 29 were detected by MS of *T. cruzi* ribosomes, including S13 and S28, which had not been detected in the proteome of *T. cruzi* [7]. Nevertheless, peptides matching to S22, S25 and S30 were not detected in the MS analysis of pure ribosomes. Interestingly, S30 was also not detected in MS studies on total extracts of *T. cruzi* (red row Table 2) [7].

For the large subunit (60S), out of the 48 yeast proteins screened, 47 were found to have a homologous gene in *T. cruzi*. The exception was L41, a short peptide of 25 amino acids long. The ribosome MS analysis detected all predicted proteins, excepting L1, L35, L39 and L40. From these, L1 and L35 were previously detected in epimastigote crude extracts [7]. Moreover, ribosome MS analysis detected two large subunit proteins that were not previously detected in the *T. cruzi* total proteome; L22 and L42 (Table 1).

In addition to the previously discussed large and small subunit proteins (S and L), MS analyses detected other well-known ribosome components. Here we discuss some examples:

i. RACK1, a protein tightly associated to the small ribosomal subunit in eukaryotes, and apparently involved in the regulation of translation initiation [9, 10]. This protein, present on the cryo-EM map of yeast ribosome was not detected on the cryo-EM map of *T. cruzi* (Figure 3). This difference can be explained by a weaker interaction between RACK1 and the ribosome in *T. cruzi*, compared to other species. This could result in too low amounts of RACK1 in purified ribosomes for cryo-EM visualization, but sufficient for MS detection.

ii. The ribosomal P proteins, a pentameric complex that form a long and protruding stalk on the large subunit involved in the translocation of the ribosome during the elongation step of protein synthesis. This complex is generally absent in cryo-EM and has not yet completely elucidated by crystallography due to its high flexibility (Figure 3). All proteins that form the P complex in *T. cruzi* were detected in purified ribosome particles, indicating the presence of a functional pentameric complex. A complete description of *T. cruzi* stalk region, components, interactions and complex formation will be discussed below.

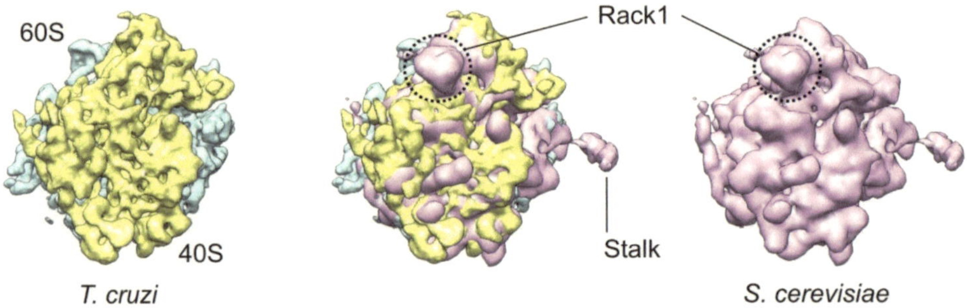

Figure 3. Left, Cryo-EM of the *T. cruzi* 80S ribosome. Center, superposition of *T. cruzi* and *S. cerevisiae* 80S particles. Right, Cryo-EM of the *S. cereviseae* 80S ribosome. Rack1 and the stalk region, which are only present in *S.cerevisiae* ribosome are indicated.

2.3. The ribosomal stalk: variable components and assembly

The large subunit of the eukaryotic ribosome possesses a long and protruding stalk formed by the ribosomal P proteins. This structure is involved in the translocation of the ribosome during the elongation step of protein synthesis through interaction with the elongation factor 2 (EF-2) [11]. Although the elongation step is a highly conserved process, the stalk is one of the most variable regions of the ribosome. The proteins forming this structure in prokaryotes and eukaryotes show very low sequence similarity. Moreover, among eukaryotes, the composition of the stalk is also variable, due to gene duplications and sequence divergence of genes encoding P proteins. In general terms the eukaryote complex is formed by the ribosomal P proteins, including P0 (a 34 kDa polypeptide) as a central component of the stalk and two distinct but closely related proteins of about 11 kDa, P1 and P2. The number of ribosomal P1/P2 proteins varies among species and these variations have consequences in the stalk composition. In mammals, the P1 and P2 families have only one member and the stalk is formed by two identical copies of each P1 and P2 proteins, linked to P0 [12]. The binding of P2 protein to P0 can only be detected in the presence of P1, suggesting a pivotal role of the latter in the conformation of the stalk [13]. In *S. cerevisiae* there are two P1 (α and β) and two P2 (α and β) proteins [14] and the stalk seems to be organized in preferential pairs; P1α/P2β and P1β/P2α. Again, both P1 proteins seem to be necessary for the binding of the corresponding P2 partners to P0 [15]. In *T. cruzi*, five components of the stalk have been identified: P0, of approximately 34 kDa, containing a C-terminal end that deviates from the eukaryotic P consensus and bears similarity with Archaea L10 protein; and four proteins of about 11 kDa (P1α, P1β, P2α and P2β) with the typical eukaryotic P consensus sequence at their C-terminal end [16, 17]. It should be noted that independent gene duplication events have originated the P1 and P2 subtypes in yeast and trypanosomatids.

Combining yeast two-hybrid technique and surface plasmon resonance (SPR) it was possible to make a complete interaction map of *T. cruzi* ribosomal P proteins [17-19]. These two techniques were both necessary to fully characterize the complex and to map the interaction regions in P0 (Figure 5). This analysis exposed some trypanosomatid-specific features among P proteins interactions. TcP0 protein (the central component of the stalk) was able to interact with all P1/P2 proteins. Both P1 proteins were unable to interact with each other nor to homo-oligomerize. Interestingly, both P2 proteins showed a highly redundant interaction profile but P2α was not able to interact with P1α. Therefore, if we focus in a *T. cruzi* stalk composed of five different ribosomal P proteins the only possible arrangement for the low molecular weight protein association will be: P1α-P2β/P1β-P2α. Any other possible combination will exclude one of the four components. This association pattern resembles that observed in yeast; although not completely, because of the presence of highly interacting components like P2β. Consequently, it is possible to postulate heterogeneity on the stalk composition in *T. cruzi*, due to redundancy of some of its components.

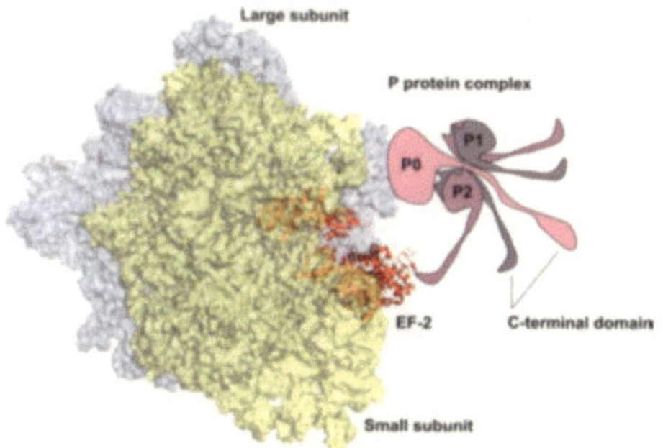

Figure 4. Surface image of yeast 80S ribosome crystallography in complex with the EF- 2 (red). The postulated location of the ribosomal P proteins on the stalk region of the large subunit is illustrated.

Since all four small P proteins can bind to P0 the question was whether TcP0 can simultaneously bind two or four proteins. Despite the accumulated data about stalk organization in several model organisms, the interaction among P0 and P1/P2 proteins is not completely understood at the molecular level. Tsurugi and Mitsui [20], based on sequence analysis of *S. cerevisiae* and *H. sapiens* P0, proposed the presence of eight putative hydrophobic zippers involved in the interaction with P1 and P2 proteins. Functional complementation assays in yeast using C-terminal truncated P0 variants showed that deletion of 87 amino acids (the entire putative hydrophobic zipper) abolishes the binding of P1/P2 proteins to the ribosome [21]. Using yeast two-hybrid technique (Y2H), it has been shown that the 100 amino acid long C-terminal domain of P0 strongly interacts with P1 and P2 proteins [22]. However, the last 50 amino acids of P0 (including only two of the eight hydrophobic residues) were not able to interact with P1/P2. More recently, using Y2H and deletion mutant strains, the P0 region between positions 213 and 260 has been involved in the interaction with the P1/P2 proteins in *S. cerevisiae*. In contrast, mutation of the putative interacting leucine residues in this region did not impair the binding of P1 and P2 proteins [23]. Based on the crystal structure of the archaeon *Pyrococcus horikoshii* stalk complex recently reported [24], it was possible to identify two putative P1/P2 interaction sites in TcP0. The first site is situated between aminoacids 222 and 232 and the second between aminoacids 248 and 258. Although the general mechanism mediating the interaction between P0 and P1/P2 proteins seems to be similar among all species, it should be noted that some of the residues on TcP0 involved in this interaction are not strictly conserved. This result is in agreement with other studies showing that ribosome from yeast strains carrying heterologous P0 proteins are not able to bind *S. cerevisiae* P1/P2 proteins efficiently [25, 26]. Altogether our data indicate that TcP0 possesses two P1/P2 interaction sites and that P1/P2 proteins can associate in pairs (P1α-P2β/P1β-P2α) but it was not known whether a hierarchy for P1/P2 association to TcP0 exists. To answer this question we performed a sequential SPR analysis in which we randomly injected one protein after the other (P1/P2) without regeneration steps on a sensor chip containing TcP0 [19]. This study showed that it is possible to form stable pentameric complexes when any of both P1 proteins were first injected. There were a few other combinations that raised stable complexes but in general terms it is possible to conclude that the injection of multi-interacting proteins (like P2β and to a lesser extent P2α) at the beginning blocks the binding of the other components of the complex. This also means that other complexes containing not all P1/P2 proteins are possible. Unfortunately, there are no functional data available.

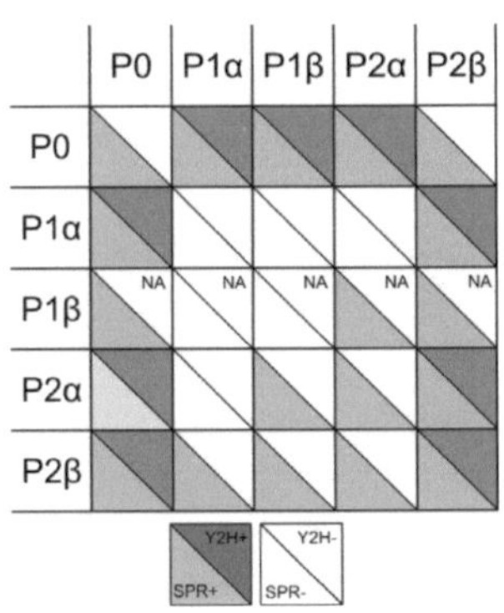

Figure 5. Summary of P protein interactions assessed by SPR and yeast two hybrid technique (Y2H). NA: Not analyzed.

Finally, the complete picture of the system can be illustrated in Figure 6, where the *T. cruzi* stalk resembles that of yeast due to the ability of all P1/P2 proteins to interact with P0.

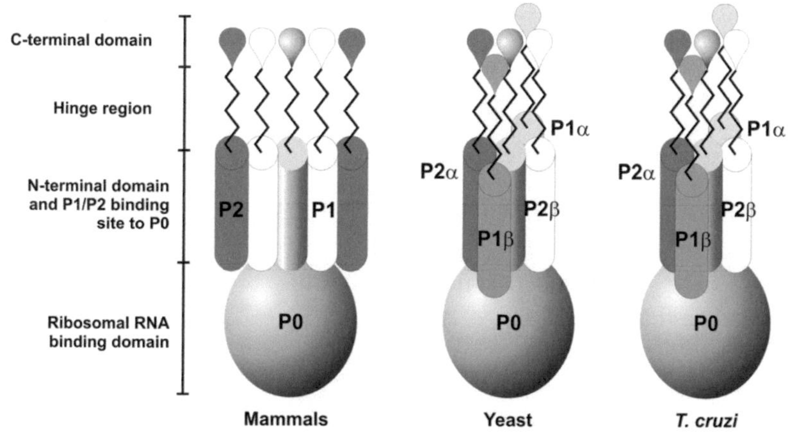

Figure 6. Assembly of the P complex in mammals, yeast and the proposed model for *T.cruzi* complex.

As it was mentioned before, the C-terminal end of ribosomal P proteins interacts with EF-2 and this interaction is essential during the elongation step of protein synthesis. Notably, antibodies against the C-terminal end of *T. cruzi* ribosomal P proteins are present in the sera of a high percentage of chronic chagasic patients. These antibodies are specific for the *T. cruzi* C-terminal peptide of ribosomal P proteins, being unable to recognize the mammalian epitope. This specificity is due to only one amino acid change (Ser by Glu) [27-30]. In a previous work, we have obtained a recombinant single chain antibody (scFvC5), derived from a monoclonal antibody against the C-terminal region of *T. cruzi* P2β protein [31, 32]. This recombinant antibody, similarly to human antibodies from chagasic patients, shows very high selectivity toward the parasite epitope. In Western blot assays, scFvC5 specifically recognized P proteins on extracts of trypanosomatids *T. cruzi*. *T. brucei* and *Crithidia fascilculata,* but it did not detect their rat counterparts. Based on earlier reports showing that antibodies (and their recombinant single chain versions) directed against the mammalian C-terminal end of P proteins inhibit protein synthesis in cell-free systems [33], we reasoned that scFvC5 would selectively block translation process by trypanosomatid ribosomes. As expected, scFvC5 strongly inhibited the incorporation of radioactive amino acids when trypanosomatid ribosomes (from *T. cruzi*, *T. brucei* and *C. fasciculata*), but not mammalian (*Rattus norvegicus*) ribosomes were used under identical experimental conditions [30]. Moreover, the translation inhibition on trypanosomatid ribosomes could be reverted by pre-incubation of the scFvC5 with the peptide corresponding to the *T. cruzi* epitope, but not with the mammalian equivalent. Therefore, we evaluated the ability of this recombinant antibody to inhibit protein synthesis *in vivo*, by using a tightly regulated inducible expression vector in *T. brucei*. The growth of parasites was significantly delayed when the scFvC5 expression was induced; clearly showing that blocking of the C-terminal end can be used as a strategy to inhibit trypanosomatid protein synthesis *in vivo*. In addition, and taking into account that the crystal structure of the monoclonal antibody originating scFvC5 has been reported [34]

(PDB 3SGE), the antibody combining site would be an interesting starting point for designing peptide mimetics as specific inhibitors of trypanosomatid translation.

3. Functional analysis

3.1. Translation activity: Initiation and elongation factors

3.1.1. Trans-splicing of trypanosomatid mRNAs

In several organisms, a variable proportion of mRNAs are processed by a mechanism named spliced-leader (SL) trans-splicing, which transfers a short RNA sequence (the SL) from the 5′ end of a specialized non-mRNA molecule, the SL RNA, to unpaired splice-acceptor sites on pre-mRNA molecules. As a result, depending on the organism, a variable proportion of the mRNAs acquires a common 5′ sequence. The SL trans-splicing mechanism is widely and patchily distributed across phylogenetically distant organisms. The evolutionary origin of this process is still an enigma, and two different hypotheses have been postulated [35]:

i. SL trans-splicing was present in an ancestral eukaryotic organism and has been lost in many different lineages, or
ii. SL trans-splicing appeared independently several times during evolution of eukaryotes.

Although this point has not been solved yet, recent evidences from analysis of large ESTs and genomic databases, seem to better support the second hypothesis [36, 37].

Several different functions for the SL sequences have been reported [35], among them:

i. providing a 5′ cap structure for protein coding RNAs transcribed by RNA polymerase I
ii. Converting polycistronic transcripts into capped, monocistronic mRNAs
iii. enhancing mRNA translational efficiency

In trypanosomatids, 100 % of their mRNA is processed by SL trans-splicing, adding a 39 nt sequence. Besides the universally conserved 7-methyl guanosine cap, which is linked to the first nucleotide via a 5′-5′ triphosphate bridge, the first four nucleotides of the SL are all methylated at the ribose ring. In addition, the first (m_2^6A) and fourth (m^3U) nucleotides are methylated at the base [38] (Figure 7). This unusually modified structure, known as the cap-4, is the most highly modified cap structure of all eukaryotic cells. Due to the important role of the mRNA 5′ end during eukaryotic translation initiation, a role for the SL structure has been proposed in the process. By using cell lines of *Leishmania tarentolae* expressing modified SL sequences, it has been shown that these modifications do not affect transcription nor trans-splicing efficiency. In contrast, mutations of the SL region spanning nucleotides 10-29 decreased the methylation extent and polysome association of mRNAs, demonstrating a direct role for the cap4 methylations and/or the primary sequence of the SL in the translation process [39]. More recently, Zamudio *et al* used *Trypanosoma brucei* strains lacking one or more of the three 2′-O-ribose methyl transferases involved in the cap4 biogenesis, to specifically evaluate the role of SL methylation in the absence of sequence changes [40]. In this study, attempts to derive cells with complete loss of mRNA cap ribose methylation were unsuccessful, indicating an essential role in kinetoplastid biology. Moreover, even when cells lacking the kinetoplastid-specific ribose methylation at positions 3 and 4 were viable, they showed a decreased rate of protein synthesis, clearly showing a role for these modification (in the absence of SL sequence changes) in the translation process. The above mentioned evidences, demonstrating a direct role of hypermethylated SL in trypanosomatid protein synthesis, reinforce the hypothesis that translation initiation would show unique features in these organisms.

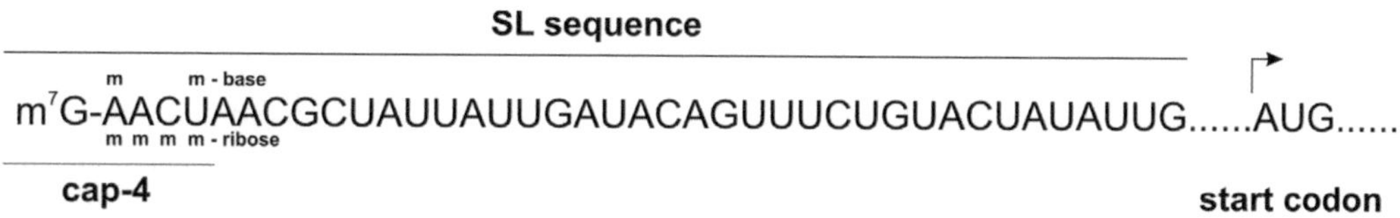

Figure 7. The *T. cruzi* spliced leader sequence

3.1.2. Initiation factors

Cap-dependent initiation in eukaryotes is a very complex, highly regulated limiting step of translation. In this process, the 5′ cap interacts with a multi-protein complex named eIF4F, formed by at least three proteins: eIF4A, an ATP-dependent RNA helicase; eIF4E, the cap-binding protein; and eIF4G, a scaffold protein that interacts with the poly(A) binding protein, IF4A and IF4E. Data mining on the *Leishmania major* genome database has revealed four eIF4Es (LmEIF4E1-4), two eIF4As (LmEIF4A1-2) and five eIF4Gs (LmEIF4G1-5) putative proteins [41].

The presence of multiple homologues for the different IF4F subunits in trypanosomatids, and in some cases in other eukaryotes, makes it difficult to identify these proteins that are actually involved in translation. In addition, knowledge of the protein synthesis in trypanosomatids is inferred by indirect evidences such as sequence similarities. Moreover, proteins with high homology to translation initiation factors are involved in other processes such as mRNA processing. Since orthologous proteins have, in general, conserved functional roles, phylogeny analysis could have some clues for identifying those proteins involved in protein synthesis. Below we discuss the biochemical, molecular and phylogenetic evidences available on the IF4F components.

3.1.3. Trypanosomatid eIF4E

A recent phylogenetic analysis of IF4E-family members has revealed that many organisms contain multiple genes encoding proteins with sequence similarity to prototypical IF4E proteins [42]. Unfortunately, no trypanosomatid IF4E-family members were included in this analysis.

The highly modified cap of trypanosomatids suggests that eIF4E orthologous in these organisms would show atypical features. The *L. major* genome revealed four putative eIF4E encoding genes (named LeishIF4E-1 to -4) with limited homology. All of them have easily identifiable orthologues in *T. brucei* and *T. cruzi*, suggesting that their functional roles are conserved among trypanosomatids, as can be inferred from phylogenetic analysis (Figure 8). In contrast, the phylogenetic relationships between putative trypanosomadid IF4E proteins and the homologous translation factors characterized in other species seem to be much more complex. This complex evolutionary scenario strongly suggests that multiple gene duplications have taken place during evolution of IF4E homologous genes, some of them before, and others after the evolutionary divergence between organisms, yielding both paralogous and orthologous genes. In spite of this complexity, it is clear that trypanosomatid IF4E-2 and IF4E-3 are paralogous genes originated by duplications into the trypanosomatid lineage.

The *L. major* genes have been cloned and their respective protein products have been biochemically characterized [43]. The corresponding recombinant proteins showed variable relative affinities for chemically synthesized mammalian and trypanosomatid cap structures. Despite a detailed study of these proteins, including pull-down assays with a mammalian interacting partner of eIF4E, analysis of co-sedimentation with polysome fractions and detection at different stages of the parasite, no clear conclusions could be obtained leading to an eIF4E candidate. Moreover, none of these proteins was able to rescue the phenotype of a *S. cerevisiae* null strain, indicating a significant functional divergence of these proteins in the trypanosomatid lineage. This conclusion is reinforced by the observation that several eIF4E homologues from phylogenetically distant organisms (mammals, *Drosophila melanogaster*, zebrafish and *Arabidopsis*) are functional in yeast (red letters in Figure 8) [44-47]. Interestingly, other IF4E homologous from these species are not functional in yeast, suggesting that they are involved in roles other than translation initiation (i.e. translation repression, binding to nuclear mRNAs).

More recently, the orthologue genes from *T. brucei* (TbEIF4E-1 through 4) have been functionally analyzed *in vivo* [50]. By using RNAi knock down of different TbIF4E genes, combined with metabolic labeling with radioactive amino acids, this study strongly suggested that TbIF4E-3 (and also probably TbIF4E-4) are directly involved in protein synthesis. This is supported by their cytosolic localization, in contrast to the TbIF4E-1 and 2 proteins, which are both nuclear and cytosolic. Unfortunately, no yeast complementation assays were performed with *T. brucei* proteins which could be correlated with the data of their *L. major* orthologues.

A similar situation seems to take place in the ancient eukaryotic, non-kinetoplastid, *Giardia lamblia*. Two IF4E homologous proteins have been identified and characterized; Gl_IF4E-1 and Gl_IF4E-2. Notably, Gl_IF4E-1 seems to be the orthologous gene of LmIF4E-4, whereas Gl_IF4E-2 is more closely related to LmIF4E-1. This suggests that duplications of genes encoding IF4E-family proteins have taken place very early during eukaryotic evolution. Functional analyses suggest that Gl_IF4E-2 is involved in protein synthesis. However, similarly to the *Leishmania* proteins, none of these IF4E homologues is able to rescue a null phenotype in yeast [51]. These observations suggest that early eukaryotes, such as *Giardia* and trypanosomatids, have major functional differences with model organisms in protein synthesis initiation, particularly in the case of IF4E. Moreover, it seems that after divergence

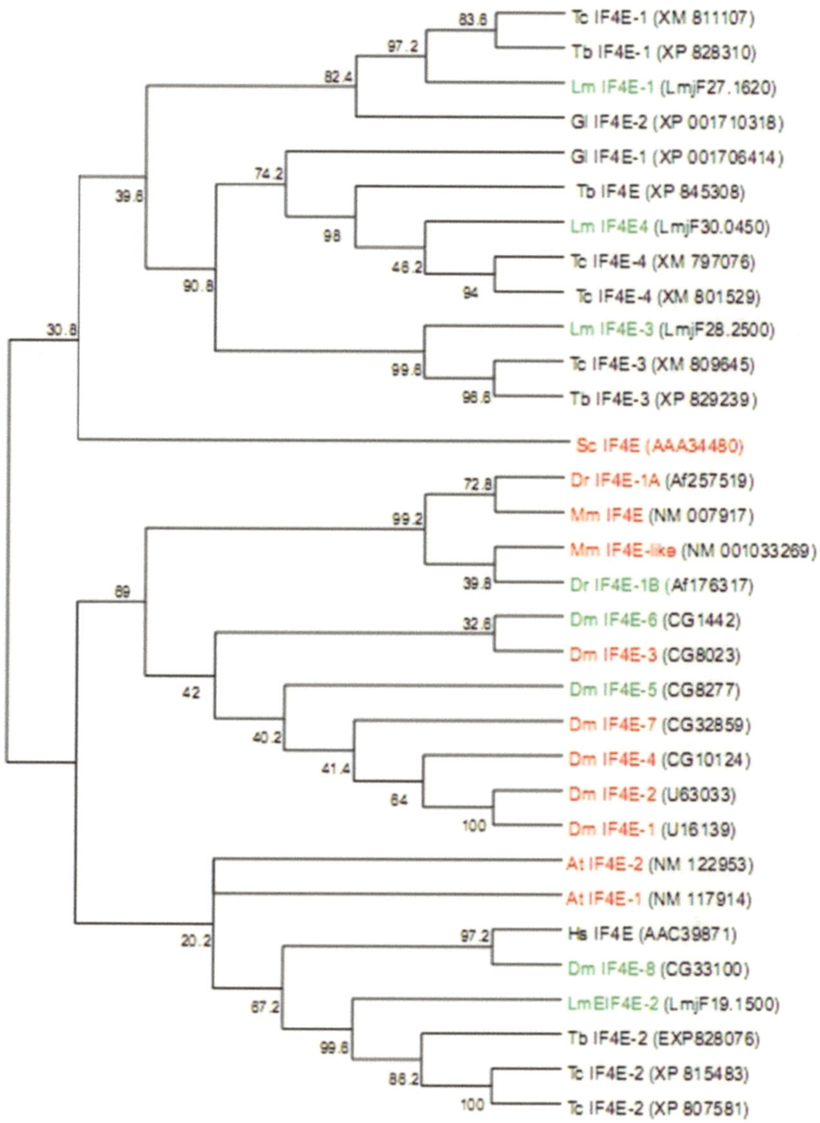

Figure 8. Inferred phylogenetic relationships among putative trypanosomatid IF4E proteins and homologous proteins from other eukaryotes. Sequences are derived from *H. sapiens* (Hs), *Mus musculus* (Mm), *S. cerevisiae* (Sc), *Drosophila melanogaster* (Dm), *Giardia lamblia*, (Gl), *Arabidopsis thaliana* (At), *Danio rerio* (zebra fish) (Dr), *L. major* (Lm), *T. brucei* (Tb) and *T. cruzi* (Tc). Accession codes for GenBank or *L. major* database (http://www.genedb.org/Homepage/Lmajor) are shown in parenthesis. Sequences covering the IF4E domain (PFAM 01652) were aligned using T-COFEE under default parameters (http://tcoffee.vital-it.ch). LG+G was chosen as the best evolutionary model using ProtTest [48]. PhyML was run using the algorithm Subtree Pruning and Regrafting (SPR) [49] with 5 initial starting trees. To estimate the robustness of the phylogenetic inference, 500 bootstrap replicates were run. Color letters indicate whether these proteins are able (red) or unable (green) to complement a *S. cerevisiae* deficient strain. Arrows indicate the proteins that have proved functional roles in protein synthesis in ancient eukaryotes (trypanosomatid and *Giardia*).

of *Giardia* and trypanosomatid lineages, different paralogous genes acquired a major functional role in protein synthesis initiation, giving an additional level of complexity to the evolution of IF4E-family genes. Data from studies in *L. major* and *T. brucei* suggest that functional roles can be inferred by orthology into the trypanosomatid lineage, but these conclusions cannot be further extrapolated to other eukaryotic organisms.

3.1.4. Trypanosomatid IF4A

The sequence analysis of IF4A gives a more simple interpretation than IF4E, because of a lower number of homologous genes in each species, and the fact that these proteins show a higher degree of sequence conservation. IF4A belongs to a DEAD-box helicase family, which includes translation factors, as well as proteins involved in splicing. The inferred phylogenetic relationships amongst these proteins yield two clearly separate clades (Figure 9). In each of these clades, the phylogeny of the proteins is coincident with that of their organisms, strongly suggesting that a unique duplication event of an ancestor IF4A gene took place before the divergence of early eukaryotes. As can be seen, *S. cerevisiae* IF4AI (P10081.3) and *Homo sapiens* IF4A2 (Q14240.2), both having demonstrated roles in protein synthesis initiation are in the same clade. Consistently, the paralogous proteins IF4A-like FAL1 (Q12099.1) and IF4AIII (P38919.4), from *S. cerevisiae* and *H. sapiens*, respectively, have been shown to be involved in splicing, and consistently they belong to a separate clade. This analysis suggests that trypanosomatid IF4A homologous proteins of the first clade, would be expected to be involved in protein synthesis initiation.

The first functional analysis of a trypanosomatid IF4A homologue was performed with a *Leishmania infantum* protein (LeIF; XP_001462692.1), initially identified as an antigen inducing IL12 mediated immune response in infected patients [52]. This protein shows RNA-dependent ATPase and ATP-dependent RNA helicase activities and interaction with yeast eIF4G *in vitro*. However, LiIF4A does not complement a yeast eIF4A deficient strain, leading the authors to suggest that LiIF4A is indeed the orthologue of human eIF4AIII, involved in splicing. The phylogenetic analysis (Figure 9) allows us to propose an alternative hypothesis, since LiIF4A is located in the translation factor clade. We propose that this protein is indeed an initiation translation factor and that its inability to complement yeast cells reflects the evolutionary functional divergence of trypanosomatid protein synthesis. This hypothesis is supported by the finding of another gene in the genome of *L. infantum*; LiIF4A-like (XP_001470194.1), which groups with the clade of splicing factors. Additional evidences supporting this hypothesis have been provided by knock-down of the two *T. brucei* paralogous genes, combined with metabolic labeling with radioactive amino acids in silenced strains. These studies showed that knock-down of TbIF4AI, but not of TbIF4AIII, specifically decreases the rate of protein synthesis [53]. Figure 9 shows a phylogenetic tree of IF4A homologous proteins from different organisms, highlighting those with demonstrated roles in translation (blue letters) or other functions than translation (pink letters).

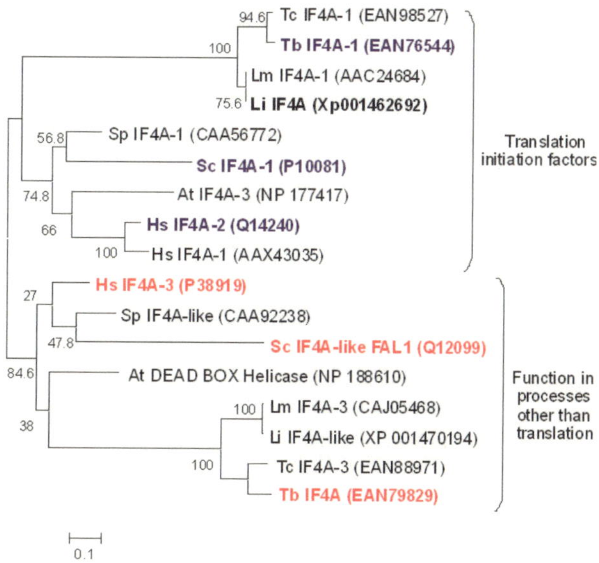

Figure 9. Inferred phylogenetic relationships among putative trypanosomatid IF4A proteins and homologous proteins from other eukaryotes. The complete open reading frames were aligned using T-COFEE under default parameters. LG+G was chosen as the best evolutionary model. All those proteins with demonstrated role in protein synthesis are located in one of these clades, whereas the second clade seems to group the proteins involved in splicing. ScIF4AI (P10081.3) [54] and HsIF4A2 (Q14240.2) [55] (blue letters), have demonstrated roles in protein synthesis, whereas ScIF4A-like FAL1 (Q12099.1) [56] and HsIF4AIII (P38919.4) [57] (pink letters), have been implicated in other processes. *L. infantum* LiIF4A (XP_001462692.1) (bold and underlined) has been cloned and characterized as unable to complement an IF4A yeast null strain [52].

3.1.5. *Trypanosomatid IF4G*

As mentioned before, IF4G is a scaffold protein that coordinates the assembly of the translation initiation factors and the 40S ribosomal subunit. The middle domain of IF4G (MIF4G) is the hallmark of these proteins. In addition, this domain is also present in several other proteins not involved in translation, such as the nuclear cap-binding protein CBP80. This fact reflects the common origin shared by IF4G with other cap-binding proteins, being difficult to deduce, based only on homology, which proteins harboring this domain have indeed initiation factor activity.

Five different proteins harboring MIF4G domain are present in the genome of *L. major*, named LmIF4G-1 to LmIF4G-5 [41]. All of them have clearly predictable orthologues in *T. brucei* and *T. cruzi*. From these, and based on different functional assays including pull-down, yeast-two hybrid and polysome profiles, LmIF4G-3 seems to be the most probable orthologous with a major role in translation initiation [41, 58]. Interestingly, phylogeny inference shows that trypanosomatid IF4G proteins form a separate clade, implying several gene duplication events into the trypanosomatid lineage (Figure 10). It has been postulated that MIF4G domain appeared during the early stages of eukaryotic evolution [59].

Interestingly, TBLASTN searches on *Giardia* genome database using many different MIF4G domains as probe, give no significant hits. This strongly suggests that this domain appeared during the evolution of eukaryotes, after the divergence of the *Giardia* lineage. This implies that trypanosomatids would be amongst the earliest diverging organisms harboring IF4G proteins.

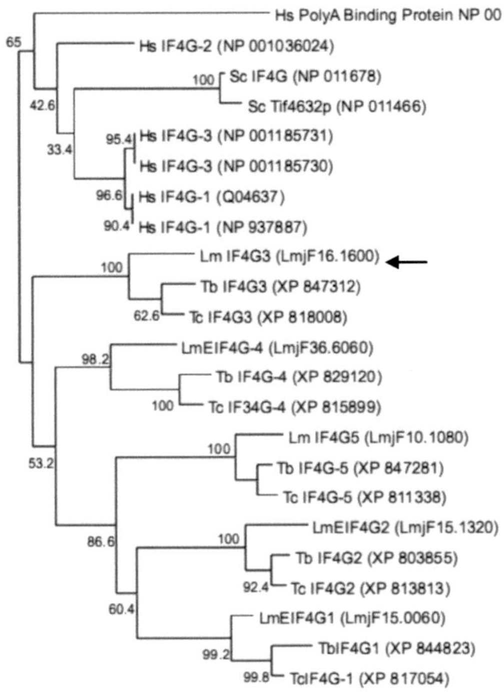

Figure 10. Inferred phylogenetic relationships among putative trypanosomatid IF4G and homologous proteins from other eukaryotes. Sequences are derived from *Homo sapiens* (Hs), *Saccharomyces cerevisiae* (Sc), *L. major* (Lm), *T. brucei* (Tb) and *T. cruzi* (Tc). Accession codes for GenBank or *L. major* database (http://www.genedb.org/Homepage/Lmajor) are shown in parenthesis. Sequences covering the MIF4G domain (PFAM cd11559) were aligned using T-COFEE under default parameters (http://tcoffee.vital-it.ch). LG + G was chosen as the best evolutionary model using ProtTest [48]. PhyML was run using the algorithm Subtree Pruning and Regrafting (SPR) [49] with 5 initial starting trees. To estimate the robustness of the phylogenetic inference, 500 bootstrap replicates were run. LmIF4G-3, the only trypanosomatid IF4G-like protein that has been experimentally associated to the formation of the IF4F complex is indicated by an arrow.

3.1.6. Elongation factors

The elongation step of protein synthesis has been highly conserved trough evolution, in comparison with initiation. According to this, the sequences encoding for elongation factors can be easily inferred by homology. However, elongation factors from bacteria and eukaryotes are not functionally interchangeable. Moreover, the specificity of ribosomes for their homologous elongation factors can be changed by interchanging their stalk components [60].

The eukaryotic EF-2 is a GTPase involved in translocation of the peptidyl-tRNA from the A site to the P site on the ribosome. Two major mechanisms involving EF-2 regulate protein elongation. One of them is EF-2 reversible phosphorylation [61]. The second is EF-2 inactivation by ADP-ribosylation of a diphthamide residue (a post-translational modification of a conserved histidine residue) [62, 63]. In addition, EF-2 interacts with the ribosomal P proteins that form the stalk region [64]. A direct interaction between the stalk base and the EF-2 has been visualized by cryoelectron microscopy in yeast [65]. Real time measurements, using surface plasmon resonance (SPR), demonstrated the binding of rat EF-2 to rat P1 and P2 proteins, with a higher affinity for P1 (K_D 3.810^{-8} M) than for P2 (K_D 2.210^{-6} M) [66]. Moreover, antibodies against the conserved C-terminal region of ribosomal P proteins inhibited the elongation step of protein synthesis, blocking the binding of EF-2 to the ribosome as well as its ribosome dependent GTPase activity [67]. Below, we will summarize evidence of structural and functional divergences amongst eukaryotes at the ribosomal stalk level, and its partner EF-2.

It has been largely demonstrated that the antifungal sordarin, which selectively inhibits some eukaryotic ribosomes, acts at the elongation step. Detailed studies about its action mechanisms have revealed that specificity of this compound for a group of fungi is due to specific molecular features of their stalk components (mainly the P0 protein) and EF-2 [68, 69].

On the other hand, trypanosomatid ribosomes have been shown to be highly resistant, in comparison to mammalian particles, to two ribosome inactivating proteins; ricin and trichosanthin [70, 71]. These toxins bind initially to the ribosomal stalk and depurinate a conserved base on the 28S rRNA, blocking the binding of EF-2 to ribosomes, supporting the idea that the trypanosomatid stalk has specific features.

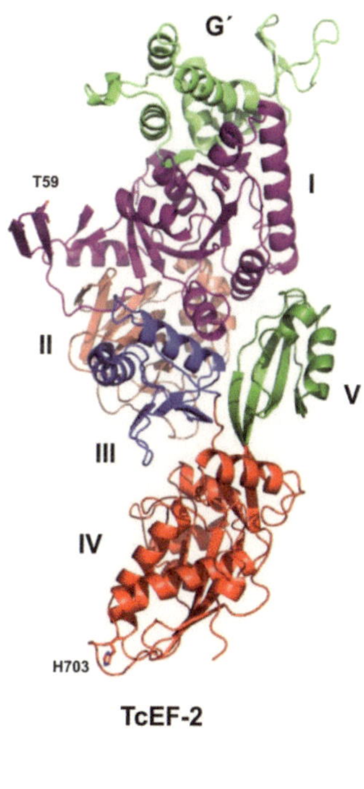

Figure 11. Model of *T. cruzi* EF-2.

The amino acid sequence of TcEF-2 was identified by probing the *T. cruzi* Gene Data Base (http://www.genedb.org) with the amino acid sequence of *S. cerevisiae* EF-2. Two identical copies of the TcEF-2 gene were found. The TcEF-2 protein was 60% identical to ScEF-2, and both EF-2 shared over 76% homology. In contrast to ScEF-2 that is regulated by phosphorylation of Threonine 57, TcEF-2 presented a Methionine in this position, suggesting a lack of regulation through phosphorylation (Figure 11). The histidine involved in ADP-ribosylation in ScEF-2 (H699) is conserved in trypanosomatids (H703 in *T. cruzi*), and consequently, diphtheria toxin inactivates protein synthesis in *T. brucei* [70]. Secondary structure comparative analysis showed an overall conserved architecture. The four canonical helices of domain I (GTPase) were conserved as well as the motif involved in nucleotide binding [72]. Using PHYRE protein fold recognition server [73], we generated a 3D model of TcEF-2 that contains the six structural domains described for ScEF-2 [74]. Therefore, taking into account the high structural similarity between ScEF-2 and TcEF-2, it may be assumed that the different domains have conserved similar functions. The interaction of TcEF-2 with the ribosomal P proteins showed that all P1/P2 proteins interacted with TcEF-2 with similar affinities. Interestingly, TcP0 showed a decreased affinity in concordance with its modified C-terminal region. Our results are in agreement with those reported by Bargis-Surgey et al. [67] for the rat homologues. However, in *T. cruzi* no differential interaction between the EF-2 and the different P1/P2 proteins was observed.

Unfortunately, data from functional comparison between trypanosomatid EF-2 and its orthologous in model species, such as yeast functional complementation, are not yet available.

3.2. Selective inhibition of trypanosomatid ribosomes by paromomycin

Aminoglycosides are a group of antibiotics binding to the decoding center of rRNA. As a result, aminoglycosides interfere with protein synthesis, facilitating amino acid misincorporation. The target site of aminoglycosides is the helix 44 of the 18S rRNA. As noted before, this helix is located in the ES12, the only expansion segment which is shorter in *T. cruzi* than in other eukaryotes, being intermediate between bacterial and higher eukaryotes [1]. Paromomycin is an aminoglycoside antibiotic with low toxicity to mammalian cells. This antibiotic has shown strong anti-leishmania activity when used alone or in combination with other drugs [75]. In a recent work, advances on the molecular basis for the differential effect of paromomycin on *Leishmania* have been reported [4]. In this work it was demonstrated that paromomycin selectively inhibits the rate of *in vitro* protein synthesis by trypanosomatid ribosomes, comparing to mammalian extracts. Moreover, the effect of paromomycin was even more dramatic when translation misreading was evaluated. Finally, affinity measurements using BIACORE demonstrated that paromomycin displays high affinity for the RNA oligonucleotide corresponding to the decoding site of trypanosomatids, whereas no interaction was detected with the oligonucleotide corresponding to the mammalian site.

4. Conclusion

In summary, we have described several direct and indirect evidences justifying future in-deep studies on the trypanosomatid protein synthesis machinery.

The sequence and functional analysis of trypanosomatid homologous to well-characterized initiation factors, along with the fact that SL sequence is present in all the trypanosomatid mRNAs, strongly suggest that protein synthesis initiation would have remarkable unique features in these protozoa. This hypothesis is reinforced by structural features of *T. cruzi* ribosomes, where a large, trypanosomatid-specific rRNA extra density is present adjacent to the platform region, involved in translation initiation.

The sequence analysis of ribosomal proteins also shows peculiarities in *T. cruzi*, being L19 and S21 proteins the most notable examples, possessing large trypanosomatid-specific domains with unknown functional roles.

Even when the elongation step of translation is much more conserved through evolution, the stalk composition and structure also seem to have special features in trypanosomatids. Consistently with this observation, some ribotoxins binding to this site show differential selectivity against parasite ribosomes. Unfortunately, no functional data are available on the functional similarities between trypanosomatid and mammalian elongation factors.

Paromomycin is the first example of a compound with a well characterized action mechanism, showing higher activity on trypanosomatid translation machinery in comparison with its mammalian counterpart. This fact, along with the number of key features of trypanosomatid ribosomes and translation factors, should strongly encourage the future search for novel, trypanosomatid-specific translation inhibitors.

Author details

Maximiliano Juri Ayub and Walter J. Lapadula
IMIBIO-SL, CONICET, Universidad Nacional de San Luis, Argentina

Johan Hoebeke
UPR9021 of the C.N.R.S. "Immunologie et Chimie Thérapeutiques", France

Cristian R. Smulski
Department of Biochemistry, University of Lausanne, Switzerland,
UPR9021 of the C.N.R.S. "Immunologie et Chimie Thérapeutiques", France

5. References

[1] Gao H, Ayub MJ, Levin MJ, Frank J. The structure of the 80S ribosome from Trypanosoma cruzi reveals unique rRNA components. Proc Natl Acad Sci U S A. 2005 Jul 19;102(29):10206-11.

[2] Frank J. Toward an understanding of the structural basis of translation. Genome Biol. 2003;4(12):237.

[3] Wuyts J, De Rijk P, Van de Peer Y, Pison G, Rousseeuw P, De Wachter R. Comparative analysis of more than 3000 sequences reveals the existence of two pseudoknots in area

V4 of eukaryotic small subunit ribosomal RNA. Nucleic Acids Res. 2000 Dec 1;28(23):4698-708.

[4] Fernandez MM, Malchiodi EL, Algranati ID. Differential effects of paromomycin on ribosomes of Leishmania mexicana and mammalian cells. Antimicrob Agents Chemother. Jan;55(1):86-93.

[5] Spahn CM, Jan E, Mulder A, Grassucci RA, Sarnow P, Frank J. Cryo-EM visualization of a viral internal ribosome entry site bound to human ribosomes: the IRES functions as an RNA-based translation factor. Cell. 2004 Aug 20;118(4):465-75.

[6] Ayub MJ, Atwood J, Nuccio A, Tarleton R, Levin MJ. Proteomic analysis of the Trypanosoma cruzi ribosomal proteins. Biochem Biophys Res Commun. 2009 Apr 24;382(1):30-4.

[7] Atwood JA, 3rd, Weatherly DB, Minning TA, Bundy B, Cavola C, Opperdoes FR, et al. The Trypanosoma cruzi proteome. Science. 2005 Jul 15;309(5733):473-6.

[8] Beckmann R, Spahn CM, Eswar N, Helmers J, Penczek PA, Sali A, et al. Architecture of the protein-conducting channel associated with the translating 80S ribosome. Cell. 2001 Nov 2;107(3):361-72.

[9] Link AJ, Eng J, Schieltz DM, Carmack E, Mize GJ, Morris DR, et al. Direct analysis of protein complexes using mass spectrometry. Nat Biotechnol. 1999 Jul;17(7):676-82.

[10] Sengupta J, Nilsson J, Gursky R, Spahn CM, Nissen P, Frank J. Identification of the versatile scaffold protein RACK1 on the eukaryotic ribosome by cryo-EM. Nat Struct Mol Biol. 2004 Oct;11(10):957-62.

[11] Liljas A. Comparative biochemistry and biophysics of ribosomal proteins. Int Rev Cytol. 1991;124:103-36.

[12] Uchiumi T, Kominami R. Binding of mammalian ribosomal protein complex P0.P1.P2 and protein L12 to the GTPase-associated domain of 28 S ribosomal RNA and effect on the accessibility to anti-28 S RNA autoantibody. J Biol Chem. 1997 Feb 7;272(6):3302-8.

[13] Gonzalo P, Lavergne JP, Reboud JP. Pivotal role of the P1 N-terminal domain in the assembly of the mammalian ribosomal stalk and in the proteosynthetic activity. J Biol Chem. 2001 Jun 8;276(23):19762-9.

[14] Planta RJ, Mager WH. The list of cytoplasmic ribosomal proteins of Saccharomyces cerevisiae. Yeast. 1998 Mar 30;14(5):471-7.

[15] Guarinos E, Remacha M, Ballesta JP. Asymmetric interactions between the acidic P1 and P2 proteins in the Saccharomyces cerevisiae ribosomal stalk. J Biol Chem. 2001 Aug 31;276(35):32474-9.

[16] Levin MJ, Vazquez M, Kaplan D, Schijman AG. The Trypanosoma cruzi ribosomal P protein family: classification and antigenicity. Parasitol Today. 1993 Oct;9(10):381-4.

[17] Juri Ayub M, Smulski CR, Nyambega B, Bercovich N, Masiga D, Vazquez MP, et al. Protein-protein interaction map of the Trypanosoma cruzi ribosomal P protein complex. Gene. 2005 Sep 12;357(2):129-36.

[18] Ayub MJ, Barroso JA, Levin MJ, Aguilar CF. Preliminary structural studies of the hydrophobic ribosomal P0 protein from Trypanosoma cruzi, a part of the P0/P1/P2 complex. Protein Pept Lett. 2005 Aug;12(6):521-5.

[19] Smulski CR, Longhi SA, Ayub MJ, Edreira MM, Simonetti L, Gomez KA, et al. Interaction map of the Trypanosoma cruzi ribosomal P protein complex (stalk) and the elongation factor 2. J Mol Recognit. 2010 Mar-Apr;24(2):359-70.

[20] Tsurugi K, Mitsui K. Bilateral hydrophobic zipper as a hypothetical structure which binds acidic ribosomal protein family together on ribosomes in yeast Saccharomyces cerevisiae. Biochem Biophys Res Commun. 1991 Feb 14;174(3):1318-23.

[21] Santos C, Ballesta JP. The highly conserved protein P0 carboxyl end is essential for ribosome activity only in the absence of proteins P1 and P2. J Biol Chem. 1995 Sep 1;270(35):20608-14.

[22] Lalioti VS, Perez-Fernandez J, Remacha M, Ballesta JP. Characterization of interaction sites in the Saccharomyces cerevisiae ribosomal stalk components. Mol Microbiol. 2002 Nov;46(3):719-29.

[23] Perez-Fernandez J, Remacha M, Ballesta JP. The acidic protein binding site is partially hidden in the free Saccharomyces cerevisiae ribosomal stalk protein P0. Biochemistry. 2005 Apr 12;44(14):5532-40.

[24] Naganuma T, Nomura N, Yao M, Mochizuki M, Uchiumi T, Tanaka I. Structural basis for translation factor recruitment to the eukaryotic/archaeal ribosomes. J Biol Chem. 2010 Feb 12;285(7):4747-56.

[25] Rodriguez-Gabriel MA, Remacha M, Ballesta JP. The RNA interacting domain but not the protein interacting domain is highly conserved in ribosomal protein P0. J Biol Chem. 2000 Jan 21;275(3):2130-6.

[26] Aruna K, Chakraborty T, Rao PN, Santos C, Ballesta JP, Sharma S. Functional complementation of yeast ribosomal P0 protein with Plasmodium falciparum P0. Gene. 2005 Aug 29;357(1):9-17.

[27] Levin MJ, Mesri E, Benarous R, Levitus G, Schijman A, Levy-Yeyati P, et al. Identification of major Trypanosoma cruzi antigenic determinants in chronic Chagas' heart disease. Am J Trop Med Hyg. 1989 Nov;41(5):530-8.

[28] Mesri EA, Levitus G, Hontebeyrie-Joskowicz M, Dighiero G, Van Regenmortel MH, Levin MJ. Major Trypanosoma cruzi antigenic determinant in Chagas' heart disease shares homology with the systemic lupus erythematosus ribosomal P protein epitope. J Clin Microbiol. 1990 Jun;28(6):1219-24.

[29] Kaplan D, Ferrari I, Bergami PL, Mahler E, Levitus G, Chiale P, et al. Antibodies to ribosomal P proteins of Trypanosoma cruzi in Chagas disease possess functional autoreactivity with heart tissue and differ from anti-P autoantibodies in lupus. Proc Natl Acad Sci U S A. 1997 Sep 16;94(19):10301-6.

[30] Juri Ayub M, Nyambega B, Simonetti L, Duffy T, Longhi SA, Gómez KA, et al. Selective blockade of trypanosomatid protein synthesis by a recombinant antibody anti-Trypanosoma cruzi P2ß proteinP2ß protein. PLoS One. 2012;7(5):e36233.

[31] Smulski C, Labovsky V, Levy G, Hontebeyrie M, Hoebeke J, Levin MJ. Structural basis of the cross-reaction between an antibody to the Trypanosoma cruzi ribosomal P2beta protein and the human beta1 adrenergic receptor. FASEB J. 2006 Jul;20(9):1396-406.

[32] Mahler E, Sepulveda P, Jeannequin O, Liegeard P, Gounon P, Wallukat G, et al. A monoclonal antibody against the immunodominant epitope of the ribosomal P2beta

protein of Trypanosoma cruzi interacts with the human beta 1-adrenergic receptor. Eur J Immunol. 2001 Jul;31(7):2210-6.

[33] Zampieri S, Mahler M, Bluthner M, Qiu Z, Malmegrim K, Ghirardello A, et al. Recombinant anti-P protein autoantibodies isolated from a human autoimmune library: reactivity, specificity and epitope recognition. Cell Mol Life Sci. 2003 Mar;60(3):588-98.

[34] Pizarro JC, Boulot G, Bentley GA, Gomez KA, Hoebeke J, Hontebeyrie M, et al. Crystal Structure of the Complex mAb 17.2 and the C-Terminal Region of Trypanosoma cruzi P2beta Protein: Implications in Cross-Reactivity. PLoS Negl Trop Dis. 2011 Nov;5(11):e1375.

[35] Hastings KE. SL trans-splicing: easy come or easy go? Trends Genet. 2005 Apr;21(4):240-7.

[36] Douris V, Telford MJ, Averof M. Evidence for multiple independent origins of trans-splicing in Metazoa. Mol Biol Evol. Mar;27(3):684-93.

[37] Derelle R, Momose T, Manuel M, Da Silva C, Wincker P, Houliston E. Convergent origins and rapid evolution of spliced leader trans-splicing in metazoa: insights from the ctenophora and hydrozoa. RNA. Apr;16(4):696-707.

[38] Bangs JD, Crain PF, Hashizume T, McCloskey JA, Boothroyd JC. Mass spectrometry of mRNA cap 4 from trypanosomatids reveals two novel nucleosides. J Biol Chem. 1992 May 15;267(14):9805-15.

[39] Zeiner GM, Sturm NR, Campbell DA. The Leishmania tarentolae spliced leader contains determinants for association with polysomes. J Biol Chem. 2003 Oct 3;278(40):38269-75.

[40] Zamudio JR, Mittra B, Campbell DA, Sturm NR. Hypermethylated cap 4 maximizes Trypanosoma brucei translation. Mol Microbiol. 2009 Jun;72(5):1100-10.

[41] Dhalia R, Reis CR, Freire ER, Rocha PO, Katz R, Muniz JR, et al. Translation initiation in Leishmania major: characterisation of multiple eIF4F subunit homologues. Mol Biochem Parasitol. 2005 Mar;140(1):23-41.

[42] Joshi B, Lee K, Maeder DL, Jagus R. Phylogenetic analysis of eIF4E-family members. BMC Evol Biol. 2005;5:48.

[43] Yoffe Y, Zuberek J, Lerer A, Lewdorowicz M, Stepinski J, Altmann M, et al. Binding specificities and potential roles of isoforms of eukaryotic initiation factor 4E in Leishmania. Eukaryot Cell. 2006 Dec;5(12):1969-79.

[44] Altmann M, Muller PP, Pelletier J, Sonenberg N, Trachsel H. A mammalian translation initiation factor can substitute for its yeast homologue in vivo. J Biol Chem. 1989 Jul 25;264(21):12145-7.

[45] Robalino J, Joshi B, Fahrenkrug SC, Jagus R. Two zebrafish eIF4E family members are differentially expressed and functionally divergent. J Biol Chem. 2004 Mar 12;279(11):10532-41.

[46] Hernandez G, Altmann M, Sierra JM, Urlaub H, Diez del Corral R, Schwartz P, et al. Functional analysis of seven genes encoding eight translation initiation factor 4E (eIF4E) isoforms in Drosophila. Mech Dev. 2005 Apr;122(4):529-43.

[47] Rodriguez CM, Freire MA, Camilleri C, Robaglia C. The Arabidopsis thaliana cDNAs coding for eIF4E and eIF(iso)4E are not functionally equivalent for yeast

complementation and are differentially expressed during plant development. Plant J. 1998 Feb;13(4):465-73.

[48] Abascal F, Zardoya R, Posada D. ProtTest: selection of best-fit models of protein evolution. Bioinformatics. 2005 May 1;21(9):2104-5.

[49] Hordijk W, Gascuel O. Improving the efficiency of SPR moves in phylogenetic tree search methods based on maximum likelihood. Bioinformatics. 2005 Dec 15;21(24):4338-47.

[50] Freire ER, Dhalia R, Moura DM, da Costa Lima TD, Lima RP, Reis CR, et al. The four trypanosomatid eIF4E homologues fall into two separate groups, with distinct features in primary sequence and biological properties. Mol Biochem Parasitol. Mar;176(1):25-36.

[51] Li L, Wang CC. Identification in the ancient protist Giardia lamblia of two eukaryotic translation initiation factor 4E homologues with distinctive functions. Eukaryot Cell. 2005 May;4(5):948-59.

[52] Barhoumi M, Tanner NK, Banroques J, Linder P, Guizani I. Leishmania infantum LeIF protein is an ATP-dependent RNA helicase and an eIF4A-like factor that inhibits translation in yeast. FEBS J. 2006 Nov;273(22):5086-100.

[53] Dhalia R, Marinsek N, Reis CR, Katz R, Muniz JR, Standart N, et al. The two eIF4A helicases in Trypanosoma brucei are functionally distinct. Nucleic Acids Res. 2006;34(9):2495-507.

[54] Blum S, Mueller M, Schmid SR, Linder P, Trachsel H. Translation in Saccharomyces cerevisiae: initiation factor 4A-dependent cell-free system. Proc Natl Acad Sci U S A. 1989 Aug;86(16):6043-6.

[55] Li W, Belsham GJ, Proud CG. Eukaryotic initiation factors 4A (eIF4A) and 4G (eIF4G) mutually interact in a 1:1 ratio in vivo. J Biol Chem. 2001 Aug 3;276(31):29111-5.

[56] Kressler D, de la Cruz J, Rojo M, Linder P. Fal1p is an essential DEAD-box protein involved in 40S-ribosomal-subunit biogenesis in Saccharomyces cerevisiae. Mol Cell Biol. 1997 Dec;17(12):7283-94.

[57] Shibuya T, Tange TO, Sonenberg N, Moore MJ. eIF4AIII binds spliced mRNA in the exon junction complex and is essential for nonsense-mediated decay. Nat Struct Mol Biol. 2004 Apr;11(4):346-51.

[58] Yoffe Y, Leger M, Zinoviev A, Zuberek J, Darzynkiewicz E, Wagner G, et al. Evolutionary changes in the Leishmania eIF4F complex involve variations in the eIF4E-eIF4G interactions. Nucleic Acids Res. 2009 Jun;37(10):3243-53.

[59] Hernandez G. On the origin of the cap-dependent initiation of translation in eukaryotes. Trends Biochem Sci. 2009 Apr;34(4):166-75.

[60] Uchiumi T, Hori K, Nomura T, Hachimori A. Replacement of L7/L12.L10 protein complex in Escherichia coli ribosomes with the eukaryotic counterpart changes the specificity of elongation factor binding. J Biol Chem. 1999 Sep 24;274(39):27578-82.

[61] Perentesis JP, Phan LD, Gleason WB, LaPorte DC, Livingston DM, Bodley JW. Saccharomyces cerevisiae elongation factor 2. Genetic cloning, characterization of expression, and G-domain modeling. J Biol Chem. 1992 Jan 15;267(2):1190-7.

[62] Ryazanov AG. Ca2+/calmodulin-dependent phosphorylation of elongation factor 2. FEBS Lett. 1987 Apr 20;214(2):331-4.

[63] Ryazanov AG, Shestakova EA, Natapov PG. Phosphorylation of elongation factor 2 by EF-2 kinase affects rate of translation. Nature. 1988 Jul 14;334(6178):170-3.

[64] Montanaro L, Sperti S, Testoni G, Mattioli A. Effect of elongation factor 2 and of adenosine diphosphate-ribosylated elongation factor 2 on translocation. Biochem J. 1976 Apr 15;156(1):15-23.

[65] MacConnell WP, Kaplan NO. The activity of the acidic phosphoproteins from the 80 S rat liver ribosome. J Biol Chem. 1982 May 25;257(10):5359-66.

[66] Gomez-Lorenzo MG, Spahn CM, Agrawal RK, Grassucci RA, Penczek P, Chakraburtty K, et al. Three-dimensional cryo-electron microscopy localization of EF2 in the Saccharomyces cerevisiae 80S ribosome at 17.5 A resolution. EMBO J. 2000 Jun 1;19(11):2710-8.

[67] Bargis-Surgey P, Lavergne JP, Gonzalo P, Vard C, Filhol-Cochet O, Reboud JP. Interaction of elongation factor eEF-2 with ribosomal P proteins. Eur J Biochem. 1999 Jun;262(2):606-11.

[68] Shastry M, Nielsen J, Ku T, Hsu MJ, Liberator P, Anderson J, et al. Species-specific inhibition of fungal protein synthesis by sordarin: identification of a sordarin-specificity region in eukaryotic elongation factor 2. Microbiology. 2001 Feb;147(Pt 2):383-90.

[69] Santos C, Rodriguez-Gabriel MA, Remacha M, Ballesta JP. Ribosomal P0 protein domain involved in selectivity of antifungal sordarin derivatives. Antimicrob Agents Chemother. 2004 Aug;48(8):2930-6.

[70] Scory S, Steverding D. Differential toxicity of ricin and diphtheria toxin for bloodstream forms of Trypanosoma brucei. Mol Biochem Parasitol. 1997 Dec 1;90(1):289-95.

[71] Juri Ayub M, Ma KW, Shaw PC, Wong KB. Trypanosoma cruzi: high ribosomal resistance to trichosanthin inactivation. Exp Parasitol. 2008 Mar;118(3):442-7.

[72] Sprang SR. G protein mechanisms: insights from structural analysis. Annu Rev Biochem. 1997;66:639-78.

[73] Kelley LA, Sternberg MJ. Protein structure prediction on the Web: a case study using the Phyre server. Nat Protoc. 2009;4(3):363-71.

[74] Jorgensen R, Ortiz PA, Carr-Schmid A, Nissen P, Kinzy TG, Andersen GR. Two crystal structures demonstrate large conformational changes in the eukaryotic ribosomal translocase. Nat Struct Biol. 2003 May;10(5):379-85.

[75] Mishra J, Saxena A, Singh S. Chemotherapy of leishmaniasis: past, present and future. Curr Med Chem. 2007;14(10):1153-69.

Section 2

Evolution and Protein Synthesis

Chapter 2

On the Emergence and Evolution of the Eukaryotic Translation Apparatus

Greco Hernández

Additional information is available at the end of the chapter

http://dx.doi.org/10.5772/48568

1. Introduction

Proteins are one of the elementary components of life and account for a large fraction of mass in the biosphere. They catalyze the big majority of reactions sustaining life, and play structural, transport, and regulatory roles in all living organisms. Hence, "Translation", i.e. the process of decoding a messenger (m)RNA by the ribosome to synthesize a protein, is a fundamental process for all forms of life (*1, 2*). Accordingly, many mechanisms to control gene expression at the translational level have evolved. They allow organisms to *i*) rapidly and reversibly respond to different stresses or sudden environmental changes; *ii*) quickly produce proteins in tissues and developmental processes where transcription is absent or limited; and *iii*) elicit asymmetric localization of proteins when is required (*1-3*). For instance, gametogenesis, early embryogenesis, memory and neurogenesis are processes where translational control plays a prominent role (*4-8*).

The knowledge of the basic processes of translation was established some decades ago, and many regulatory mechanisms have been subsequently elucidated in different organisms (*9-11*). In recent years, the use of powerful genome-wide sequencing, proteomics and bioinformatics-based technologies in both model and non-model organisms, has shown that a number of components of the translation apparatus has undergone a large diversification across eukaryotes, and that many of them emerged at different times on evolution (*11, 12*). Moreover, universal and lineage-specific mechanisms regulating translation have been described, and evidence supports the notion that some of them might have emerged at different times during evolution (*11, 12*). Evidence supports the notion that some of these mechanisms might have appeared by tinkering, i.e. co-opting and assembling molecules and regulatory mechanisms from other cellular processes (*11*). Overall, the emerging view suggests two general principles. On one hand, that while the fundaments of translation are well conserved across all forms of life, in eukaryotes the initiation step has undergone substantial increase in complexity as compared to prokaryotes (*11, 13-18*); on the other hand, that after eukaryotes emerged the translation apparatus continued evolving to certain degree during eukaryotic diversification (*12*). The continue divergence of eukaryotes led to the diversification of metabolic requirements, to the appearance of different levels of body plans and organismal complexity, and to the arousal of novel developmental programs and behavioural patterns. Altogether, these changes led to the invasion of novel ecological niches. These events most probably were both the causes and effects of a parallel diversification and specialization, to different levels in different taxa, of components and mechanisms of the translation apparatus. Here I will review current knowledge on how this apparatus might have originated and further evolved in eukaryotes, making emphasis on the initiation step of translation.

2. An overview of the translation process in eukaryotes

Eukaryotic translation is a sophisticated and tightly regulated process, the basic steps of which are conserved in all eukaryotes. It is performed by the ribosome together with multiple 'translation' factors and aminoacyl-tRNA synthetases (aaRSs). It is divided into four major groups of steps: initiation, elongation, termination and recycling (Fig. 1).

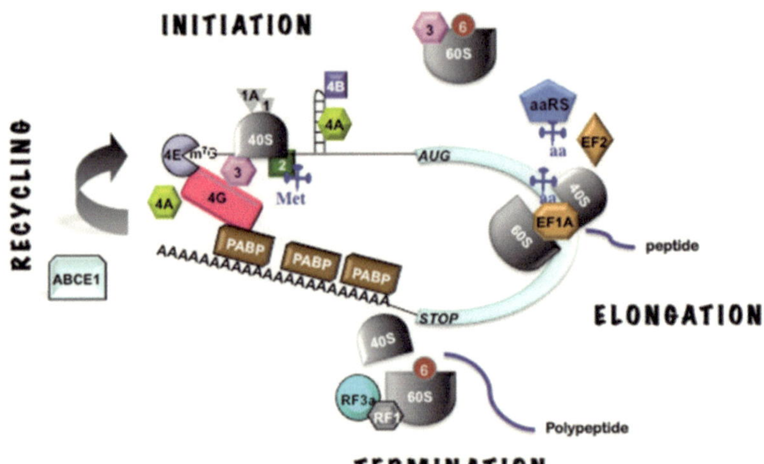

Figure 1. The general process of translation in eukaryotes. A typical eukaryotic mRNA is represented. The cap structure (m^7G), the open reading frame (light blue box) and the poly(A) tail are depicted. During *Initiation*, most eukaryotic mRNAs are translated by the cap-dependent mechanism, which requires recognition by eIF4E (light purple) complexed with eIF4G (red) and eIF4A (light green) –the so-called eIF4F complex– of the cap structure at the 5′ end. A 43S pre-initiation complex (consisting in a 40S ribosomal subunit (dark gray) loaded with eIF3 (pink), eIF1 and eIF1A (light grey), initiator Met-$tRNA_i^{Met}$ (blue clover), eIF2 (dark green) and GTP binds the eIF4F-mRNA complex and scans along the 5′-UTR of the mRNA to reach the start codon (usually an *AUG* triplet). During the scanning eIF4A, stimulated by eIF4B (dark blue), unwinds secondary RNA structure in an ATP-dependent manner. The poly A-binding protein (PABP, dark brown) binds both the poly(A) tail and eIF4G promoting mRNA circularization. Free 60S ribosomal subunit is stabilized by eIF6 and eIF3. *Elongation* is assisted by elongation factors eEF1A and eEF2 (light brown). During this step, aminoacyl-tRNA synthetases (aaRSs,

blue) catalyze the binding of amino acids (*aa*) to cognate tRNAs. *Termination* is mediated by the release factors eRF1 (gray) and eRF3 (light blue), and happens when a termination codon (*STOP*) of the mRNA is exposed in the A-site of the ribosome. In this step, the completed polypeptide (blue line) is released. During *Recycling*, which is required to allow further rounds of translation, both ribosomal subunits dissociate from the mRNA. Recycling is assisted by ABCE1 (light blue). eRF1 remains associated with the post-termination complexes after polypeptide release.

2.1. Initiation

Translation initiation is mediated by eukaryotic initiation factors (eIF). For the big majority of mRNAs, translation initiation happens by the so-called cap-dependent mechanism (*19-23*). It begins with the dissociation of the ribosome into its 60S and 40S subunits by eIF6. Free 40S subunit, which is stabilized by eIF3, eIF1 and eIF1A, binds to a ternary complex (consisting of eIF2 bound to an initiator Met-tRNA$_i^{Met}$ and GTP) to form a 43S pre-initiation complex. In separate events, the cap structure (m^{7}GpppN, where N is any nucleotide) present in the 5′ end of the mRNA is recognized by eIF4E in complex with eIF4G (forming the so-called eIF4F complex). Then the 43S pre-initiation complex is recruited to the 5′ end of the mRNA, a process that is coordinated by eIF4G through its interactions with eIF4E and the 40S ribosomal subunit-associated eIF3. eIF4G also interacts with the poly A-binding protein (PABP) which interacts with the mRNA 3′-poly(A) tail, thereby promoting circularization of the mRNA and increasing its stability. The closed-loop model proposes that during translation, cross-talk occurs between both mRNA ends due to this circularity. The ribosomal complex then scans in a 5′ —> 3′ direction along the 5′-untranslated region (UTR) of the mRNA to reach the start codon, usually an AUG. During scanning, eIF4B stimulates the activity of eIF4A, which unwinds secondary RNA structures in the mRNA. eIF1, eIF1A, and eIF5 assist in the positioning of the 40S ribosomal subunit at the correct start codon so that eIF2 can deliver the anti-codon of the initiator Met-tRNA$_i^{Met}$ as the cognate partner for the start codon, directly to the peptidyl (P)-site of the 40S ribosomal subunit. Once the ribosomal subunit is placed on the start codon, a 48S pre-initiation complex is formed. Then eIF5 promotes GTP hydrolysis by eIF2 to release the eIFs. Finally, the 60S ribosomal subunit joins the 40S subunit in a eIF5B-dependent manner to form an 80S initiation complex. The outcome of the initiation process is a 80S ribosomal complex assembled at the start codon of the mRNA containing a Met-tRNA$_i^{Met}$ in the P-site (*19-22*).

In some mRNAs, 5′-UTR recognition by the 40S ribosomal subunit is driven by RNA structures located in *cis* within the mRNA itself. Such structures are defined as Internal Ribosome Entry Site (IRES) and are located nearby the start codon. This mechanism takes place without involvement of the cap structure and eIF4E and is called an IRES-dependent initiation of translation (*13, 24, 25*).

2.2. Elongation

Translation elongation is assisted by elongation factors (eEF). During this step, mRNA codons are decoded and peptide bonds are formed sequentially to add amino acid residues to the carboxy-terminal end of the nascent, mRNA-encoded, peptide (*21, 26-28*). Elongation involves four major steps: *i*) formation of the ternary complex eEF1A·GTP·aminoacyl-tRNA and delivery of the first elongator aminoacyl-tRNAs to an empty ribosomal tRNA-binding site called the A (acceptor)-site. It is in the A-site where codon/anticodon decoding takes place; *ii*) Interaction of the ribosome with the mRNA-tRNA. This duplex activates eEF1A·GTP hydrolysis and guanine nucleotide exchange on eEF1A; *iii*) Peptide bond formation then occurs between the P-site peptidyl-tRNA and the incoming aminoacyl moiety of an A-site aminoacyl-tRNA. This reaction is catalyzed by the peptidyl transferase center of the 60S ribosomal subunit, and the products comprise of a new peptidyl-tRNA that is one amino acid residue longer and a deacylated (discharged) tRNA. *iv*) Binding of eEF2·GTP and GTP hydrolysis promote the translocation of the mRNA such that the deacylated tRNA moves to the E (exit)-site, the peptidyl-tRNA is in the P-site, and the mRNA moves by three nucleotides to place the next mRNA codon into the A-site. The deacylated tRNA in E-site is then ejected from the ribosome. The whole process is repeated along the mRNA sequence. When a stop codon is reached the process of termination is initiated (*21, 26-28*).

2.3. Termination

Translation termination is mediated by two polypeptide chain-release factors, eRF1 and eRF3. When any of the termination codons is exposed in the A-site, eRF1 recognizes the codon, binds the A-site, and triggers the release of the nascent polypeptide from the ribosome by hydrolysing the ester bond linking the polypeptide chain to the P-site tRNA. This reaction leaves the P-site tRNA in a deacylated state, leaving it to be catalyzed by the peptidyl transferase center of the ribosome. eRF1 recognizes stop signals and functionally acts as a tRNA-mimic, whereas eRF3 is a ribosome- and eRF1-dependent GTPase that, by forming a stable complex with eRF1, stimulates the termination process (*21, 29, 30*).

2.4. Recycling

In the recycling step, both ribosomal subunits are dissociated releasing the mRNA and deacetylated tRNA, so that both ribosomal subunits can be used for another round of initiation (*21, 29, 30*) (Fig.1). Evidence suggests that the ABC-type ATPase ABCE1 is probably the general ribosome recycling factor which coordinates termination with recycling (*31*). ABCE1 establishes multiple contacts with both ribosomal subunits as well as with the release factors and stimulates ribosome dissociation. ABCE1 also influences eRF1 function during stop-codon recognition and peptidyl-tRNA hydrolysis. During ribosome recycling, eRF3 dissociates from ribosomal complexes after GTP hydrolysis, whereas eRF1 remains associated with posttermination ribosomal complexes after peptide release (*31-33*).

According to the closed-loop model, termination and recycling may not release the 40S ribosomal subunit back into the cytoplasm. Instead, this subunit may be passed from the poly(A) tail back to the 5′-end of the mRNA, so that a new round of initiation can be started on the same mRNA (*21, 29, 34*)

3. The emergence of eukaryotic translation

The emergence of eukaryotes from prokaryotic ancestors led toward novel, higher levels of cell organization. A plethora of new features emerged at the cellular level, including the acquisition of a nucleus, an endoplasmic reticulocyte and endosymbiotic bacteria, the formation of split genes sorted out in chromosomes, an actin-based cytoskeleton and, in many phyla, the emergence of multicellularity, behavioural patterns and developmental programs. In this new type of cell, novel features appeared in the translation process, including the spatio-temporal separation between transcription and translation, and the increase of the number of events occurring during the initiation step that led to the establishment of the cap-dependent mechanism.

During the emergence of eukaryotes, the translation apparatus itself also underwent profound changes, including the evolution of the 40S and 60S ribosomal subunits from the prokaryotic 30S and 50S, respectively. This was due to the addition of several rRNA expansion segments, peptide additions to most ribosomal proteins, and the addition of extra eukaryotic-specific components, including novel ribosomal proteins and the 5.8S rRNA (*26, 35-38*). Moreover, the number of initiation factors increased. While in prokaryotes translation initiation is assisted by three factors, in eukaryotes initiation needs the interplay of at least fourteen factors. Thus, novel, eukaryotic-specific translation factors evolved (including eIF3, eIF4B, eIF4E, eIF4G, eIF4H and eIF5). The mRNA also underwent profound changes which can be summarized as follows: *i*) it acquired a novel molecular structure, from polycistronic to monocistronic, capped, and polyadenylated transcripts; *ii*) it acquired a novel life cycle, from simultaneous transcription/translation, to be transcribed, spliced and exported from the nucleus, to be stored, translated and degraded in the cytoplasm; and *iii*) it acquired a novel functional conformation when engaged in translation, displaying a functional cross-talk between the 5′- and 3′-ends. Finally, with the emergence of eukaryotes the number of different mechanisms that regulate translation was expanded (*11-15, 17, 39*).

How the transition from prokaryotic to eukaryotic translation occurred remains still unresolved. I will discuss some ideas that have been set forward to try to address this question.

4. The transition from prokaryotic to eukaryotic translation initiation

4.1. Translation initiation in the prokaryotic world

It was established in the 1970s that in eubacteria the recruitment of the small ribosomal subunit to the mRNA occurs by a direct interaction. This happens via the complementary base pairing between the Shine-Dalgarno (SD) sequence of the mRNA, which is a purine-rich region located at around 10 nucleotides upstream the start codon, and a sequence at the 3′ end of the 16S rRNA on the ribosome (referred to as anti-Shine-Dalgarno sequence, ASD) (*40, 41*). The importance of the SD sequence to initiate translation was later experimentally corroborated in different eubacteria and archaea (*42-44*), and has been retained in some cell organelles that evolved from eubacteria over a billion years ago (*45*). This, together with the large proportion of genes having the SD sequence in the well studied bacteria, led to the general idea that for the vast majority of prokaryotic mRNAs the SD sequence was the essential (although not necessarily the sole) element to select the correct initiation codon, and that the SD/ASD interaction during initiation is conserved in most prokaryotes (*16, 46, 47*). However, in recent years a large number of mRNAs lacking a SD sequence have been discovered widespread in a variety of different eubacterial and archaeal lineages. These include mRNAs devoid of 5'-UTR (and hence referred to as "leaderless" mRNAs) (*15, 16, 39, 45, 48-54*), and mRNAs that possess a 5′-UTR and lack a SD sequence (*45, 48-53, 55*). For leaderless mRNAs the start codon itself was found to serve as the most important signal for ribosome recruitment and for translation initiation. Here the initiator tRNA and IF2 are critical for complex formation between the start codon and the ribosome. It is noteworthy that translation initiation of leaderless mRNAs involves the undissociated ribosome 70S instead of the 30S ribosomal subunit (*15, 16, 39, 51, 56-59*). mRNAs with a 5′-UTR devoid of a SD sequence exhibit a pronounced minimum in secondary structure and AUG start codons reside in single-stranded regions of the mRNAs. For these mRNAs, ribosome binding to the start codon is a sequence-independent event, but is strictly dependent on the local absence of RNA secondary (*45*). Translation initiation of these transcripts is promoted by the ribosomal protein S1 (RPS1), which is a component of the 30S ribosomal subunit. RPS1 interacts with the 5′-UTR of an mRNA initiating translation efficiently, regardless of the presence of a SD sequence. Intriguingly, neither archaeobacteria nor eukaryotes contain a RPS1 gene, raising the question of how leadered mRNAs devoid of a SD sequence are translated in Archaea (*39, 46, 49, 51*). Finally, evidence suggests that alternative, unknown mechanisms might be used to initiate translation in Cyanobacteria (*49*) and in haloarchaea (*60*).

Overall, the emerging view indicates that in the prokaryotic world, both SD-dependent and SD-independent translation mechanisms are present in all major groups of prokaryotes. Indeed, evidence suggests that the leaderless mechanism might represent the major pathway to initiate translation in Archaea (*52-54*). Thus, it has been suggested that the last common ancestor of existing life already possessed an established fundamental translational apparatus, but the mechanisms to initiate translation initiation further diversified in the bacterial and archaeal lineages (*17, 49, 50, 60*).

4.2. What was the mechanism to recruit mRNAs in the last common ancestor of existing life?

Despite the presence of leaderless and leadered, SD-lacking mRNAs across prokaryotes, a recent study using the genomes of 277 prokaryote species, both eubacteria and archaea, showed that the anti-SD sequence at the 3′ end of the 16S rRNA on the ribosome is highly conserved among all species, and that loss of the SD sequence seems to have occurred multiple times, independently, in different phyla (*49*). These observations strongly suggest that the SD/ASD interaction plays an important role in translation initiation in essentially all prokaryote species that are descended from the last universal common ancestor. Thus, the SD-based mechanism of ribosome recruiting might have driven translation initiation in the last common ancestor of existing organisms, but was further lost in different phyla (*49*). For those phyla which lost the SD sequence, RSP1-mediated or leaderless mRNA-used mechanisms of translation initiation work to great extent (*49*). The evolutionary pressures that led to the loss of SD sequences, however, are completely unknown.

4.3. The transition to eukaryotes

As mentioned above, while the fundaments of translation are well conserved in all forms of life, in eukaryotes the initiation step underwent substantial increase in complexity and in number of initiation factors as compared to prokaryotes.

Although it is established that eukaryotes evolved from archaeal ancestors, we still don't know what lineage they evolved from. Thus, we don't know what type of mRNA (i.e. SD-containing, leaderless, or SD-lacking transcript) the first eukaryotes possessed. Nevertheless, all eukaryotic mRNAs lack SD sequences and ribosomes have no RPS1. I have previously suggested that three were the most important evolutionary forces that led to the emergence of the cap-dependent initiation mechanism in eukaryotes, namely (*i*) the need to adjust to the emergence of the nuclear membrane and interrupted genes, (*ii*) the subsequent requirement to splice and export intron-less mRNAs to the cytoplasm, and (*iii*) the absence of SD sequence and RSP1 in eukaryotic mRNAs (*11, 13, 14*). Because eukaryotic mRNAs lack both SD sequences and RPS1 protein, they cannot efficiently recruit the small ribosomal subunit directly to the initiation codon. This, together with the fact that most initiation factors that evolved only within the eukaryotes (including eIF4E, eIF4G, eIF4B, eIF4H and eIF3) are involved in the cap-binding and the scanning processes, indicates that the absence of both SD sequences and RSP1 protein was one of the crucial selection pressures that led to early eukaryotes to develop a novel mechanism to ensure the correct landing of the ribosome at the 5′-end of mRNAs, namely the cap-dependent initiation (*11, 13, 14*).

I have discussed that during eukaryogenesis and before the time when the cap-dependent initiation was developed, it is possible that there would have been a transition period where the mRNAs of the early eukaryotes were translated in a cap-independent, IRES-driven manner. In this period, 5′-UTRs lacking SD motifs that were able to passively recruit the 40S ribosomal subunit would have been positively selected for and could, therefore, have become the first examples of IRESs (Fig. 2). In this scenario, the cap structure, a proto-eIF4G, the poly(A) tail and an ancestral PABP, might have appeared for functions in RNA metabolism that emerged among the primary adaptive responses to the emergence of split genes and the need for nucleocytoplasmic RNA export, but initially had no role in translation (*11, 13, 14*). As a consequence of the absence of the SD sequence in eukaryotic mRNAs, the scaffold proteins eIF4G, eIF3 and eIF4B, as well as the 5′-end cap structure, eIF4E, and RNA helicases were later incorporated into the already established but incipient eukaryotic translation machinery because they ensured a more efficient recruitment of the 40S ribosomal subunit by the mRNA. Altogether, these events led the "scanning" process to evolve and to the establishment of the current cap-dependent translation initiation. Mutations in PABP, which allowed binding to eIF4G and promoted mRNA circularization underwent a strong positive selection because they ensured a more efficient recruitment of the 40S ribosomal subunit by the mRNA, stimulating translation and increasing mRNA stability. mRNA circularization provides an effective means for the protein synthesis apparatus to selectively translate only intact mRNAs, i.e., those that harbor both a cap and a poly(A) tail. Thus, mRNA circularization also (and perhaps primarily) underwent a strong positive selection because it represents a checkpoint that determines whether or not to initiate translation (*11, 13, 14*).

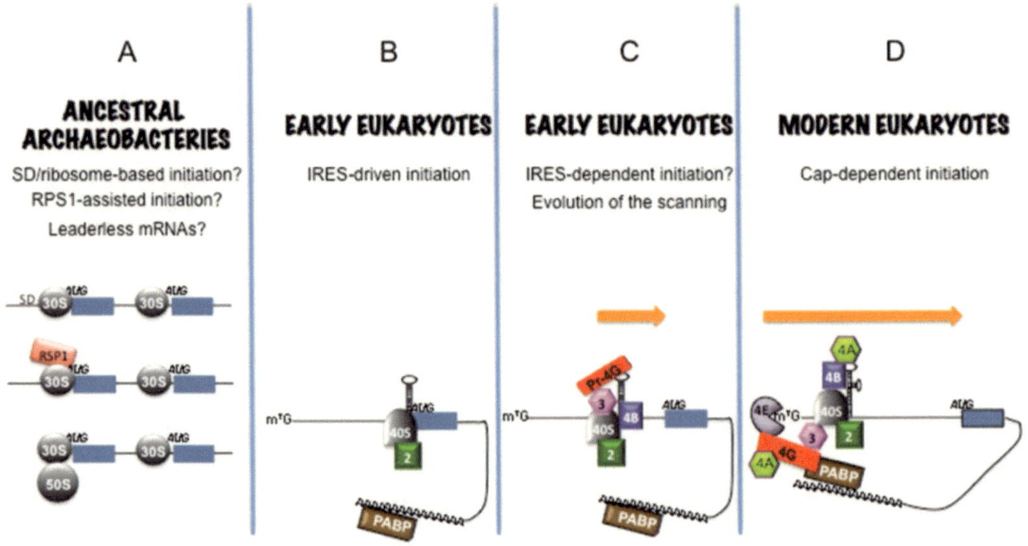

Figure 2. A proposed model for the evolution of the cap-dependent.
initiation of translation. (A) Ancestral archaeal cells had polycistronic (blue boxes) mRNAs. It is not known what was the mechanism used by the prokaryotic ancestors of eukaryotes to recruit the mRNA by the 30S ribosomal subunit. (B) The evolutionary transition from ancestral archaeal cells to early eukaryotes is represented. Due to the appearance of the nucleus and split genes, transcription and translation were decoupled, and a need to splice, export and protect the transcripts during nucleus-cytoplasm export emerged. The cap structure might have first appeared at this evolutionary stage to provide a "platform" to assemble the splicing, export and RNA protection mechanisms (see reference *14* for details). The arousal of PABP might have happened at this stage as part of the mechanisms to protect mRNAs (see reference *11* for details). The lack of SD sequences in the mRNAs, probably as a result of massive invasion of introns from endosimbionts, as well as the apparition of monocistronic mRNAs and both the 40S and the 60S ribosomal subunits, happened at this stage. Initiation of translation occurred perhaps in an IRES-dependent manner via the direct recruitment of the 40S ribosomal subunit by the mRNA. (C) The evolution in the cytoplasm of a proto-eIF4G (Pr-4G), along with the emergence of eIF3 and eIF4B, gradually improved the delivery of the 40S ribosomal subunit to the early mRNAs during evolution. The scanning process (orange arrow) evolved probably at this stage due to the activity of RNA helicases coming from different processes of metabolism that at the same time could participate in the unwinding of mRNA secondary structure during translation initiation. Among them, however, only an eIF4A-like helicase evolved functional interactions with eIF4G thus becoming later the canonical initiation factor eIF4A. (D) It is not know when eIF4E evolved during eukaryogenesis. However, an interaction of eIF4G with eIF4E, eIF4A and PABP evolved, which eventually led to the establishment of today's widespread cap-dependent mechanism to initiate translation.

5. Diversification of eukaryotes and further evolution of the translation apparatus

5.1. Functional divergence of initiation factors

After eukaryotes emerged, components of the translation machinery further evolved along the radiation into different phyla (Fig. 3). The continue appearance of different levels of organismal complexity led to the arousal of new phyla, developmental programs, behavioral patterns, and the invasion of novel ecological niches by eukaryotes. These events most probably were the causes of a further evolution, to different levels, of components and mechanisms of the translation apparatus in different taxa (*11, 12*). In the following, I will summarize the most studied examples of this.

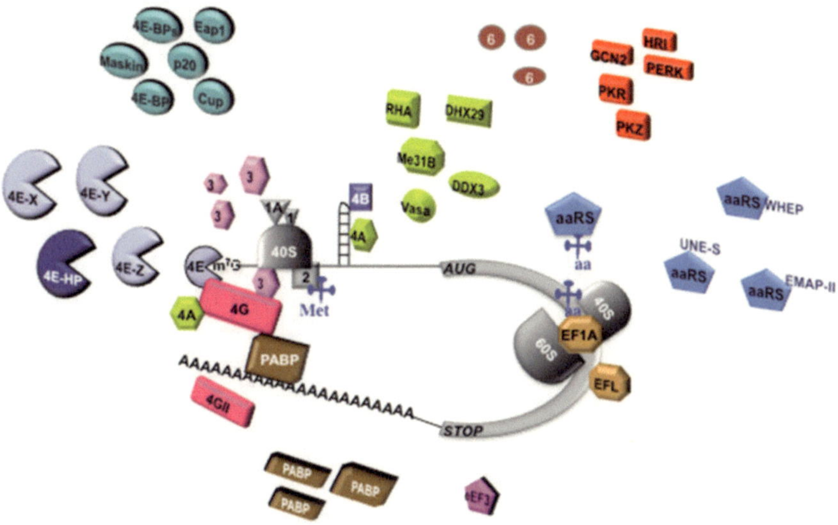

Figure 3. Diversity in the components of the translation apparatus across eukaryotes. The different components of the translation machinery with well-studied diversity in different phyla are shown in colors. Components with some diversity that is not discussed here are depicted in gray. Several cognates of eIF4E (purple) and eIF4G (red) have been found in plants, metazoan, protists and some fungi. In some cases, eIF4E cognates have evolved towards translational repressors (4E-HP, dark blue, is an example). Many 4E-binding proteins (including Maskin, 4E-BPs, Eap1, p20 and Cup, light blue) have been discovered in different species. The subunit composition of eIF3 (pink) ranges from 5 to 13 nonidentical polypeptides in different phyla. There is, however, a core of five homolog subunits shared by most eukaryotes. Some diversity has been found in eIF6. Several RNA helicases (light green) from diverse organisms are involved in the *Initiation* step. A family of five kinases (*HRI, PERK, GCN2, PKR* and *PKZ*, red) phosphorylate the alpha subunit of eIF2 to inhibit global translation under stress conditions. The presence of eIF2alpha kinases varies in different lineages. Different domains, such as *WHEP, EMAPII*, and *UNE-S*, have been added to different aminoacyl-tRNA synthetases (*aaRSs*, blue) in distinct phyla of multicellular species. For *Elongation* to happen, a number of protist, algae and fungi lack eEF1A (light brown) and instead possess the related factor elongation factor-like (EFL, dark brown). Ribosomes from all eukaryotes perform *Elongation* with eEF1A and eEF2. However, the yeast *S. cerevisiae* requires an additional essential factor, eEF3 (dark pink), for *Elongation* to proceed. Genes encoding eEF3 have been found exclusively in many species of fungi. Evidence supports the notion that eEF3 activity promotes ribosome recycling. Several cytoplasmic PABP (dark brown) cognates have been found in many phyla.

Most evidence for molecular and functional diversification among the translation components has been found in the eIF4 proteins. All major eukaryotic lineages possess several paralog genes for members of the eIF4 families. For some of them, differential expression patterns, and even variable biochemical properties among paralogs of the same organism, have been found (*12, 14, 61-66*). For eIF4E and eIF4G cognates, physiological specialization has been also found and, in some cases, eIF4E cognates have evolved towards translational repressors (*12, 14, 61-63, 65*). These findings support the hypothesis that in organisms with several paralogs, a ubiquitous set of eIF4 factors supports global translation initiation whereas other paralogs perform their activity in specific cellular processes (*61*). Whereas the need for distinct eIF4 proteins in different tissues may have been the driving force behind the evolution of various paralogs in multicellular organisms, in unicellular eukaryotes different paralogs may be differentially needed during distinct life stages (*67*).The multisubunit eIF3 is another example of factor that has undergone molecular diversification across eukaryotes, whose subunit composition ranges from 5 to 13 nonidentical polypeptides in different phyla (*68*). However, the functional relevance of these phenomena is not known.

5.2. Multiple helicases involved in the initiation step

eIF4A is the factor traditionally thought to perform RNA helicase activity to unwind the 5'-UTR secondary structure during the scanning. Recently, other RNA helicases from diverse organisms have also been found to be involved in different steps of translation initiation. Such is the case of the mammalian, *Drosophila* and yeast helicases DDX3 and Ded1, as well as the human helicases RHA and DHX29 (*69-71*). Evidence supports the idea that in *Drosophila,* the helicase Vasa is a translational activator of specific mRNAs involved in germline development (*5, 6*). In contrast, orthologs of the *Xenopus* helicase Xp54 in several organisms, including *Drosophila* Me31B, *Saccharomyces* Dhh1, human rck/p54, and *Caenorhabditis* CGH-1 have been found to repress translation of stored mRNAs and promote aggregation into germplasm-containing structures (*72*). In *Arabidopsis,* the eIF4F complex contains eIF4A in proliferating cells but different RNA helicases in quiescent cells (*73*). These findings show that evolutionary convergence has happened in different lineages to fulfill the need of RNA helicase activity during translation initiation.

5.3. Divergence in molecules involved in the elongation step

In contrast to the initiation step, the process of elongation is highly conserved among all forms of life. Strikingly, a recent genome-wide analysis revealed that a number of eukaryotic lineages lack eEF1A, a canonical factor that delivers aa-tRNAs to the A-site of ribosomes during the elongation step. Instead, they possess a related factor called elongation factor-like (EFL) protein that retains the residues critical for eEF1A (*74*). It was later found that EFL-encoding species are scattered widely across eukaryotes and that *eEF1A* and *EFL* genes display mutually exclusive phylogenetic distributions. Thus, it is assumed that eEF1A and EFL are functionally equivalent (*74-82*). It is thought that *eEF1A* is ancestral to all extant eukaryotes and that a single duplication event in a specific lineage gave rise to *EFL*. *EFL* genes were then spread to other lineages via multiple independent lateral gene transfer events, where EFL took over the original eEF1A function resulting in secondary loss of the endogenous *eEF1A*. It is thought that both genes co-existed for some time before one or the other was lost. Indeed, the diatom *Thalassiosira* bears both *EFL* and *eEF1A* genes (*79*) and might be an example of this situation. It is also possible that there was a single gain of *EFL* early in evolution followed by differential loss of it (*74, 78, 79, 81, 82*). So far, *EFL* genes have been identified in widespread taxa, including diatoms, green and red algae, fungi, euglenozoans, foraminiferans, cryptophytes, goniomonads, katablepharid, chlorarachniophytes, oomycetes, dinoflagellates, choanozoans, centrohelids and haptophytes. Most of them are unicellular organisms. In contrast, most eukaryotes contain only *eEF1A* (*74-82*).

Key molecules for elongation are aaRSs, which catalyze the aminoacylation reaction whereby an amino acid is attached to the cognate tRNA (*21, 26, 27*). aaRS represent an intriguing and perhaps unique case of evolution among the components of the translation apparatus. Throughout evolution of multicellularity, different domains such as the tripeptide ELR (Glu-Leu-Arg), the oligonucleotide binding fold-containing EMAPII domain, the WHEP domain, the glutathione S-transferase (GST) domain and a specialized amino-terminal helix (N-helix), have been progressively added to different aaRSs in distinct phyla. The tripeptide ELR and the EMAPII domain were incorporated simultaneously to TyrRSs in metazoans starting from insects; the WHEP domain is present in TrpRS only in chordates; and a unique sequence motif, UNE-S, became fused to the C-terminal of SerRS of all vertebrates. In bilaterian the glutamylRS and prolylRS were linked via WHEP domains giving rise to a bifunctional glutamyl-prolylRS (*83, 84*).

It has been found that the function of the aaRSs was either increased or impaired by the addition of the new domains. Whereas the WHEP domain regulates interaction of TrpRS with its cognate receptor, with MetRS this domain plays a tRNA-sequestering function. The Leu-zipper motif in ArgRS is important for the formation of multi aaRSs complex (MSC), which enhances channeling of tRNA to the ribosome. Moreover, different aaRSs play diverse roles in cellular activities beyond translation, such as stress response, plant and animal embryogenesis, cell death, immune responses, transcriptional regulation, and RNA splicing (*83-85*). It has been found that the incorporation of domains to aaRSs correlates positively with the increase in organism's complexity. For example, the number of aaRSs carrying the GST domain increases from two in fungi to four in insects, to five in fish and six in humans (*83*). Thus, it has been proposed that the newly fused aaRSs domains triggered the appearance of new biological functions for these proteins in different lineages, and that the fusion of domains to aaRSs could have played an important part in expanding the complexity of newly emerging metazoan phyla (*83*).

5.4. Divergence in termination and recycling factors

Termination is governed by eRF1, which is a monophyletic protein that was inherited by eukaryotes from archaeal ancestors. eRF1 is universally present in eukaryotes and, with the exception of some vascular plants and some ciliates, eukaryotes contain only one *eRF1* gene (*86-89*). Interestingly, unusually high rates of eRF1 evolution have been found in organisms with variant genetic codes (mostly protists and unicellular fungi), especially in the N-terminal domain, which is responsible for stop-codon recognition (*86, 87, 89-92*). eRF1 displays structural similarity to tRNA molecules and mimics its activity during binding of ribosomal A-site during recognition of a stop codon (*91-94*). Since mutations in eRF1 N-terminal domain switch from omnipotent to bipotent mode for stop-codon specificity (*94-98*), most likely the accelerated evolution of eRF1 in organisms with variations to the nuclear genetic code has been driven mainly to accommodate these variations (*89-99*).

Another striking case of evolutionary divergence was found in the fungi. Ribosomes from all eukaryotes perform elongation with eEF1A and eEF2. Surprisingly, it was found that the yeast *Saccharomyces cerevisiae* requires an additional essential factor, eEF3, for the elongation cycle to proceed (*100*). Genes encoding eEF3 were subsequently identified exclusively in other fungi, including *Candida, Pneumocystis, Neurospora, Aspergillus* and *Mucor* (*101-104*). eEF3 is an ATPase that interacts with both ribosomal subunits and that is required for the binding of aminoacyl-tRNA-eEF1A-GTP ternary complex to the ribosomal A-site by enhancing the rate of deacylated tRNA dissociation from the E-site (*105*). Recently, it was shown that post-termination complex, consisting of a ribosome, mRNA, and tRNA, is disassembled into single components by ATP and eEF3. Because the release of mRNA and deacylated tRNA and ribosome dissociation takes place simultaneously and no 40S—mRNA complexes remain, it is proposed that eEF3 activity promotes ribosome recycling (*106*). It remains unsolved what were the evolutionary forces that led to the emergence of eEF3 exclusively in fungi.

6. Concluding remarks

One of the enigmas in current Biology is how eukaryotic protein synthesis emerged. I have discussed that, in the absence of SD sequence in mRNAs and RPS1 in ribosomes, the evolution of translation machinery followed a gradual addition of scaffold proteins, namely eIF4G, eIF3 and eIF4B, which highly improved the efficiency and regulation of mRNA binding to the 40S ribosomal subunit (*14*). This, together with the incorporation of several RNA helicases, eIF4E and PABP, gradually improved the global efficiency, accuracy and possibilities of gene expression regulation at the level of translation initiation (*14*). Most likely the molecular diversification of the translation apparatus is among the basis that provided to early eukaryotes the scope to invade new ecological niches and overcome the different environmental and biological challenges this represented. Indeed, the evolution of the translation apparatus might have been both, cause and consequence of eukaryotic radiation.

Translation in eukaryotes is tightly coupled to other features and components of cellular metabolism. For example, translation control is coupled to RNA transport to ensure different developmental programs to occur (*107, 108*). The RNA transport machineries have also diverged in different phyla, and together with them some components of the translation apparatus also diverged (*108*). Another fundamental aspect of RNA metabolism is the storage and degradation of mRNAs in different cytoplasmic bodies, such as P-bodies and P-granules, which contain translation factors (*109*). The diversity and conservation of these foci across phyla are a reflection of the general evolution that the translation machinery and its regulation have undergone during eukaryotes evolution.

Author details

Greco Hernández
Division of Basic Research, National Institute for Cancer (INCan), Tlalpan, Mexico City, Mexico

Acknowledgement

I was supported by the National Institute for Cancer (Instituto Nacional de Cancerología, México).

7. References

[1] Mathews, M. B., Sonenberg, N., and Hershey, J. W. B. (2000) Origins and principles of translational control, in *Translational control of gene expression* (Sonenberg, N., Hershey, J. W. B., and Mathews, M. B., Eds.), pp 1-31, Cold Spring Harbor Laboratory press, Cold Spring Harbor, New York.

[2] Mathews, M. B., Sonenberg, N., and Hershey, J. W. B. (2007) Origins and principles of translational control, in *Translational control in biology and medicine* (Mathews, M. B., Sonenberg, N., and Hershey, J. W. B., Eds.), pp 1-40, Cold Spring Harbor Laboratory Press, Cold Spring Harbor, New York.

[3] Mazumder, B., Seshadri, V., and Fox, P. L. (2003) Translational control by the 3'-UTR: the ends specify the means. *Trends Biochem. Sci. 28*, 91-98.

[4] Renkawitz-Pohl, R., Hempel, L., Hollman, M., and Schafer, M. A. (2005) Spermatogenesis, in *Comprehensive molecular insect science* (Gilbert, L. I., Iatrou, K., and Gill, S. S., Eds.), pp 157-177, Elsevier Pergamon.

[5] Richter, J. D., and Lasko, P. (2011) Translational control in oocyte development, *Cold Spring Harbor Perspect. Biol. 3*, a002758.

[6] Lasko, P. (2009) Translational control during early development, in *Progress in Molecular Biology and Translational Science* (Hershey, J. W. B., Ed.), pp 211-254, Academic Press, Burlington.

[7] Thompson, B., Wickens, M., and Kimble, J. (2007) Translational control in development, in *Translational control in biology and medicine* (Mathews, M. B., Sonenberg, N., and Hershey, J. W. B., Eds.), pp 507-544, Cold Spring Harbor Laboratory Press, Cold Spring Harbor, New York.

[8] Sonenberg, N., and Hinnebusch, A. G. (2007) New modes of translation control in development, behavior, and disease, *Mol. Cell 28*, 721-729.

[9] Mathews, M. B., Sonenberg, N., and Hershey, J. W. B., (Eds.) (2007) *Translational control in biology and medicine,* Cold Spring Harbor Laboratory Press, Cold Spring Harbor, New York.

[10] Sonenberg, N., Hershey, J. W. B., and Mathews, M. B., (Eds.) (2000) *Translational control of gene expression,* Cold Spring Harbor Laboratory press, Cold Spring Harbor, New York.

[11] Hernández, G., Altmann, M., and Lasko, P. (2010) Origins and evolution of the mechanisms regulating translation initiation in eukaryotes. *Trends Biochem. Sci. 35,* 63-73.

[12] Hernández, G., Proud, C. G., Preiss, T., and Parsyan, A. (2012) On the diversification of the translation apparatus across eukaryotes. *Comp. Funct. Genom.* In press.

[13] Hernández, G. (2008) Was the initiation of translation in early eukaryotes IRES-driven? *Trends Biochem. Sci. 33,* 58-64.

[14] Hernández, G. (2009) On the origin of the cap-dependent initiation of translation in eukaryotes. *Trends Biochem. Sci. 34,* 166-175.

[15] Benelli, D., and Londei, P. (2009) Begin at the beginning: evolution of translational initiation. *Res. Microbiol. 160,* 493-501.

[16] Londei, P. (2005) Evolution of translational initiation: news insights from the archaea. *FEMS Microbiol. Rev. 29,* 185-200.

[17] Kyrpides, N. C., and Woese, C. R. (1998) Universally conserved translation initiation factors. *Proc. Natl. Acad. Sci. U. S. A. 95,* 224-228.

[18] Aravind, L., and Koonin, E. V. (2000) Eukaryotic-specific domains in translation initiation factors: implications for translation regulation and evolution of the translation system. *Genome Res. 10,* 1172-1184.

[19] Hershey, J. W. B., and Merrick, W. C. (2000) Pathway and mechanism of initiation of protein synthesis, in *Translational control of gene expression* (Sonenberg, N., Hershey, J. W. B., and Mathews, M. B., Eds.), pp 33-88, Cold Spring Harbor Laboratory press, Cold Spring Harbor, New York.

[20] Jackson, R. J., Hellen, C. U., and Pestova, T. V. (2010) The mechanism of eukaryotic translation initiation and principles of its regulation. *Nat. Rev. Mol. Cell Biol. 11,* 113-127.

[21] Kapp, L. D., and Lorsch, J. R. (2004) The molecular mechanics of eukaryotic translation. *Annu. Rev. Biochem 73,* 657-704.

[22] Sonenberg, N., and Hinnebusch, A. (2009) Regulation of translation initiation in eukaryotes: mechanisms and biological targets. *Cell 136,* 731-745.

[23] Hinnebusch, A. G. (2011) Molecular mechanism of scanning and start codon selection in eukaryotes. *Microbiol. Mol. Biol. Rev. 75,* 434-467.

[24] Martínez-Salas, E., Piñeiro, D., and Fernández, N. (2012) Alternative mechanisms to initiate translation in eukaryotic mRNAs. *Comp. Funct. Genom.* In press.

[25] Elroy-Stein, O., and Merrick, W. C. (2007) Translation initiation by viral cellular internal ribosome entry sites, in *Translational control in biology and medicine* (Mathews, M. B., Sonenberg, N., and Hershey, J. W. B., Eds.) 2nd ed., pp 155-172, Cold Spring Harbor Laboratory Press, Cold Spring Harbor, New York.

[26] Taylor, D. J., Frank, J., and Kinzy, T. G. (2007) Structure and function of the eukaryotic ribosome and elongation factors, in *Translational control in biology and medicine* (Mathews, M. B., Sonenberg, N., and Hershey, J. W. B., Eds.), Cold Spring Harbor Laboratory Press, Cold Spring Harbor, New York.

[27] Andersen, G. R., Nissen, P., and Nyborg, J. (2003) Elongation factors in protein synthesis. *Trends Biochem. Sci. 28*, 434-441.

[28] Herbert, T. P., and Proud, C. G. (2007) Regulation of Translation Elongation and the Cotranslational Protein Targeting Pathway, in *Translational control in biology and medicine* (Mathews, M. B., Sonenberg, N., and Hershey, J. W. B., Eds.), pp 601-624, Cold Spring Harbor Laboratory Press, Cold Spring Harbor, New York.

[29] Ehrenberg, M., Hauryliuk, V., Crist, C. G., and Nakamura, Y. (2007) Translation termination, the prion [PSI+], and ribosomal recycling, in *Translational control in biology and medicine* (Mathews, M. B., Sonenberg, N., and Hershey, J. W. B., Eds.), pp 173-196, Cold Spring Harbor Laboratory Press, Cold Spring Harbor, New York.

[30] Jackson, R. J., Hellen, C. U. T., and Pestova, T. V. (2012) Termination and post-termination events in eukaryotic translation. *Advances Prot. Chem. Struct. Biol. 86*, 45-93.

[31] Becker, T., Franckenberg, S., Wickles, S., Shoemaker, C. J., Anger, A. M., Armache, J. P., Sieber, H., Ungewickell, C., Berninghausen, O., Daberkow, I., Karcher, A., Thomm, M., Hopfner, K. P., Green, R., and Beckmann, R. (2012) Structural basis of highly conserved ribosome recycling in eukaryotes and archaea. *Nature 482*, 501-506.

[32] Pisarev, A. V., Hellen, C. U., and Pestova, T. V. (2007) Recycling of eukaryotic posttermination ribosomal complexes. *Cell 131*, 286-299.

[33] Pisarev, A. V., Skabkin, M. A., Pisareva, V. P., Skabkina, O. V., Rakotondrafara, A. M., Hentze, M. W., Hellen, C. U., and Pestova, T. V. (2010) The role of ABCE1 in eukaryotic posttermination ribosomal recycling. *Mol. Cell 37*, 196-210.

[34] Jacobson, A. (1996) Poly(A) metabolism and translation: the closed-loop model, in *Translational control* (Hershey, J. W. B., Mathews, M. B., and Sonenberg, N., Eds.), pp 451-480, Cold Spring Harbor Laboratory Press.

[35] Klinge, S., Voigst-Hoffmann, F., Leibundgut, M., and Ban, N. (2012) Atomic structures of the eukaryotic ribosome. *Trends Biochem. Sci.* In press.

[36] Dresios, J., Panopoulos, P., and Synetos, D. (2006) Eukaryotic ribosomal proteins lacking a eubacterial counterpart: important players in ribosomal function. *Mol. Microbiol. 59*, 1651-1663.

[37] Yokoyama, T., and Suzuki, T. (2008) Ribosomal RNAs are tolerant toward genetic insetions: evolutionary origin of the expansion segments. *Nucleic Acid Res. 36*, 3539-3551.

[38] Hartman, H., Favaretto, P., and Smith, T. F. (2006) The archaeal origins of the eukaryotic translational system. *Archaea 2*, 1-9.

[39] Benelli, D., and Londei, P. (2011) Translation initiation in Archaea: conserved and domain-specific features. *Biochem. Soc. Trans. 19*, 89-93.

[40] Shine, J., and Dalgarno, L. (1974) The 3'-terminal sequence of *Escherichia coli* 16S ribosomal RNA: complementary to nonsense triplets and ribosome binding sites. *Proc. Natl. Acad. Sci. U.S.A. 71*, 1342-1346.

[41] Steitz, J. A., and Jakes, K. (1975) How ribosomes select initiator regions in mRNA: base-pair formation between the 3' terminus of 16s rRNA and the mRNA during initiation of protein synthesis in *Escherichia coli. Proc. Natl. Acad. Sci. U.S.A. 72*, 4734-4738.
[42] Jacob, W. F., Santer, M., and Dahlberg, A. E. (1987) A single base change in the Shine-Dalgarno region of 16S rRNA of *Escherichia coli* affects translation of many proteins. *Proc. Natl. Acad. Sci. U.S.A. 84*, 4757-4761.
[43] Band, L., and Henner, D. J. (1984) *Bacillus subtilis* requires a "stringent" Shine-Dalgarno region for gene expression. *DNA 3*, 17-21.
[44] Dennis, P. P. (1997) Ancient ciphers: translation in archaea. *Cell 89*, 1007-1010.
[45] Scharff, L. B., Childs, L., Walther, D., and Bock, R. (2011) Local absence of secondary structure permits translation of mRNAs that lack ribosome-binding site,.*PLoS Genet. 7*, e1002155.
[46] Jackson, R. J. (2000) A comparative view of initiation site selection mechanisms, in *Translational control of gene expression* (Sonenberg, N., Hershey, J. W. B., and Mathews, M. B., Eds.), pp 127-183, Cold Spring Harbor Laboratory press, Cold Spring Harbor, New York.
[47] Laursen, B. S., Sørensen, H. P., Mortensen, K. K., and Sperling-Petersen, H. U. (2005) Initiation of protein synthesis in bacteria. *Microbiol. Mol. Biol. Rev. 69*, 101-123.
[48] Chang, B., Halgamuge, S., and tang, S. L. (2006) Analysis of SD sequences in completed microbial genomes: non-SD-led genes are as common as SD-led genes. *Gene 373*, 90-99.
[49] Nakagawa, S., Niimura, Y., Miura, K. I., and Gojobori, T. (2010) Dynamic evolution of translation initiation mechanisms in prokaryotes. *Proc. Natl. Acad. Sci. U. S. A. 107*, 6382-6387.
[50] Zheng, X., Hu, G. Q., She, Z. S., and Zhu, H. (2011) Leaderless genes in bacteria: clue to the evolution of translation initiation mechanisms in prokaryotes. *BMC Genomics 12*, 361-370.
[51] Moll, I., Grill, S., Gualerzi, C. O., and Blasi, U. (2002) Leaderless mRNAs in bacteria: surprises in ribosomal recruitment and translational control. *Mol. Microbiol. 43*, 239-246.
[52] Brenneis, M., Hering, O., Lange, C., and Soppa, J. (2007) Experimental characterization of Cis-acting elements important for translation and transcription in halophilic archaea. *PLoS Genet. 3*, e229.
[53] Slupska, M. M., King, A. G., Fitz-Gibbon, S., Besemer, J., Borodovsky, M., and Miller, J. H. (2001) Leaderless transcripts of the crenarchaeal hyperthermophile *Pyrobaculum aerophilum. J. Mol. Biol. 309*, 347-360.
[54] Tolstrup, N., Sensen, C. W., Garrett, R. A., and Clausen, I. G. (2000) Two different and highly organized mechanisms of translation initiation in the archaeon *Sulfolobus solfataricus. Extremophiles 4*, 175-179.
[55] Torarinsson, E., Klenk, H. P., and Garret, R. A. (2005) Divergent transcriptional and translational signals in Archaea. *Environ. Microbiol. 7*, 45-54.
[56] Grill, S., Gualerzi, C. O., Londei, P., and Bläsi, U. (2000) Selective stimulation of translation of leaderless mRNA by initiation factor 2: evolutionary implications for translation. *EMBO J. 19*, 4101-4110.

[57] O'Donnell, S. M., and Janssen, G. R. (2001) The initiation codon affects ribosome binding and translational efficiency in *Escherichia coli* of cI mRNA with or without the 5' untranslated leader. *J. Bacteriol. 183,* 1277-1283.
[58] O'Donnell, S. M., and Janssen, G. R. (2002) Leaderless mRNAs bind 70S ribosomes more strongly than 30S ribosomal subunits in *Escherichia coli. J. Bacteriol. 184,* 6730-6733.
[59] Moll, I., Hirokawa, G., Kiel, M. C., A., K., and Bläsi, U. (2004) Translation initiation with 70S ribosomes: an alternative pathway for leaderless mRNAs. *Nucleic Acid Res. 32,* 3354-3363.
[60] Hering, O., Brenneis, M., Beer, J., Suess, B., and Soppa, J. (2009) A novel mechanism for translation initiation operates in haloarchaea. *Mol. Microbiol. 71,* 1451-1463.
[61] Hernández, G., and Vazquez-Pianzola, P. (2005) Functional diversity of the eukaryotic translation initiation factors belonging to eIF4 families. *Mech. Dev. 122,* 865-876.
[62] Jagus, R., Bachvaroff, T. R., Joshi, B., and Place, A. R. (2012) Diversity of eukaryotic translation initiation factor eIF4E in protists. *Comp. Funct. Genom.* In press.
[63] Patrick, R. M., and Browning, K. S. (2012) The eIF4F and eIFiso4F complexes of plants: an evolutionary perspective. *Comp. Funct. Genom. 2012,* In press.
[64] Joshi, B., Lee, K., Maeder, D. L., and Jagus, R. (2005) Phylogenetic analysis of eIF4E-family members. *BMC Evol. Biol. 5,* 48.
[65] Rhoads, R. E. (2009) eIF4E - new family members, new binding partners, new roles. *J. Biol. Chem. 284,* 16711-16715.
[66] Zinoviev, A., and Shapira, M. (2012) Evolutionary conservation and diversification of the translation initiation apparatus in trypanosomatids. *Comp. Funct. Genom. 2012,* In press.
[67] Freire, E. R., Dhalia, R., Moura, D. M., Lima, T. D., Lima, R. P., Reis, C. R., Hughes, K., Figueiredo, R. C., Standart, N., Carrington, M., and de Melo, N. O. P. (2011) The four trypanosomatid eIF4E homologues fall into two separate groups, with distinct features in primary sequence and biological properties. *Mol. Biochem. Parasitol. 176,* 25-36.
[68] Hinnebusch, A. G. (2006) eIF3: a versatile scaffold for translation initiation complexes. *Trends Biochem. Sci. 31,* 553-560.
[69] Parsyan, A., Svitkin, Y., Shahbazian, D., Gkogkas, C., Lasko, P., Merrick, W. C., and Sonenberg, N. (2011) mRNA helicases: the tacticians of translational control. *Nature Rev. Mol. Cell Biol. 14,* 235-245.
[70] Stevenson, A. L., and McCarthy, J. E. G. (2008) Found in translation: another RNA helicase function. *Mol. Cell 32,* 755-756.
[71] Linder, P., and Jankowsky, E. (2011) From unwinding to clamping - the DEAD box RNA helicase family. *Nature Rev. Mol. Cell Biol. 12,* 505-516.
[72] Weston, A., and Sommerville, J. (2006) Xp54 and related (DDX6-like) RNa helicases: roles in messenger RNP assembly, translation regulation and RNA degradation. *Nucleic Acids Res. 34,* 3082-3094.
[73] Bush, M. S., Hutchins, A. P., Jones, A. M. E., Naldrett, M. J., Jarmolowski, A., Lloyd, C. W., and Doonan, J. H. (2009) Selective recruitment of proteins to 5' cap complexes during he growth cycle in *Arabidosis. Plant J. 59,* 400-412.

[74] Keeling, P. J., and Inagaki, Y. (2004) A class of eukaryotic GTPase with a punctate distribution suggesting multiple functional replacements of translation elongation factor 1alpha. *Proc. Natl. Acad. Sci. U. S. A. 101*, 15380-15385.

[75] Sakaguchi, M., Takishita, K., Matsumoto, T., Hashimoto, T., and Inagaki, Y. (2009) Tracing back EFL gene evolution in the cryptomonads-haptophytes assemblage: separate origins of EFL genes in haptophytes, photosynthetic cryptomonads, and goniomonads. *Gene 441*, 126-131.

[76] Gile, G. H., Novis, P. M., Cragg, D. S., Zuccarello, G. C., and Keeling, P. J. (2009) The distribution of Elongation Factor-1 Alpha (EF1alpha), Elongation Factor-Like (EFL), and a non-canonical genetic code in the ulvophyceae: discrete genetic characters support a consistent phylogenetic framework. *J. Eukar. Microbiol. 56*, 367-372.

[77] Cocquyt, E., Verbruggen, H., Leliaert, F., Zechman, F. W., Sabbe, K., and De Clerck, O. (2009) Gain and loss of elongation factor genes in green algae. *BMC Evol. Biol. 9*, 39.

[78] Noble, G. P., Rogers, M. B., and Keeling, P. J. (2007) Complex distribution of EFL and EF1alpha proteins in the green algal lineage, *BMC Evolutionary Biology 7*, 82.

[79] Kamikawa, R., Inagaki, Y., and Sako, Y. (2008) Direct phylogenetic evidence for lateral transfer of elongation factor-like gene. *Proc. Natl. Acad. Sci. U. S. A. 105*, 6965-6969.

[80] Kamikawa, R., Yabuki, A., Nakayama, T., Ishida, K., Hashimoto, T., and Inagaki, Y. (2011) Cercozoa comprises both EF1α-containing and EFL-containing members. *Eur. J. Protistol. 47*, 24-28.

[81] Kamikawa, R., Sakaguchi, M., Matsumoto, T., Hashimoto, T., and Inagaki, Y. (2010) Rooting for the root of elongation factor-like protein phylogeny. *Mol. Phylogenet. Evol. 56*, 1082-1088.

[82] Gile, G. H., Faktorová, D., Castlejohn, C. A., Burger, G., Lang, B. F., Farmer, M. A., Lukes, J., and Keeling, P. J. (2009) Distribution and phylogeny of EFL and EF1alpha in Euglenozoa suggest ancestral co-occurrence followed by differential loss, *PLoS ONE 4*, e5162.

[83] Guo, M., Yang, X. L., and Schimmel, P. (2010) New functions of aminoacyl-tRNA synthetases beyond translation. *Nature Rev. Mol. Cell Biol. 11*, 668-674.

[84] Szymański, M., Deniziak, M., and Barciszewski, J. (2000) The new aspects of aminoacyl-tRNA synthetases. *Acta Biochim. Polonica 47*, 821-834.

[85] Fox, P. L., Ray, P. S., Arif, A., and Jia, J. (2007) Noncanonical Functions of Aminoacyl-tRNA Synthetases in Translational Control, in *Translational control in biology and medicine* (Mathews, M. B., Sonenberg, N., and Hershey, J. W. B., Eds.), pp 829-854, Cold Spring Harbor Laboratory Press, Cold Spring Harbor, New York.

[86] Kim, O. T., Yura, K., Go, N., and Harumoto, T. (2005) Newly sequenced eRF1s from ciliates: the diversity of stop codon usage and the molecular surfaces that are important for stop codon interactions. *Gene 346*, 277-286.

[87] Moreira, D., Kervestin, S., Jean-Jean, O., and Philippe, H. (2002) Evolution of eukaryotic translation elongation and termination factors: variations of evolutionary rate and genetic code deviations. *Mol. Biol. Evol. 19*, 189-200.

[88] Atkinson, G. C., Baldauf, S. L., and Hauryliuk, V. (2008) Evolution of nonstop, no-go and nonsense-mediated mRNA decay and their termination factor-derived components. *BMC Evol. Biol. 8*, 290-308.

[89] Inagaki, Y., and Doolittle, W. F. (2001) Class I release factors in ciliates with variant genetic codes. *Nucleic Acids Res. 29*, 921-927.

[90] Knight, R. D., Freeland, S. J., and Landweber, L. F. (2001) Rewiring the keyboard: evolvability of the genetic code. *Nature Rev. Genet. 2*, 49-58.

[91] Lozupone, C. A., Knight, R. D., and Landweber, L. F. (2001) The molecular basis of nuclear genetic code change in ciliates. *Curr. Biol. 11*, 65-74.

[92] Song, H., Mugnier, P., Das, A. K., Webb, H. M., Evans, D. R., Tuite, M. F., Hemmings, B. A., and Barford, D. (2000) The crystal structure of human eukaryotic release factor eRF1-mechanism of stop codon recognition and peptidyl-tRNA hydrolysis. *Cell 100*, 311-321.

[93] Kolosov, P., Frolova, L., Seit-Nebi, A., Dubovaya, V., Kononenko, A., Oparina, N., Justesen, J., Efimov, A., and Kisselev, L. (2005) Invariant amino acids essential for decoding function of polypeptide release factor eRF1. *Nucleic Acids Res. 33*, 6418-6425.

[94] Ito, K., Frolova, L., Seit-Nebi, A., Karamyshev, A., Kisselev, L., and Nakamura, Y. (2002) Omnipotent decoding potential resides in eukaryotic translation termination factor eRF1 of variant-code organisms and is modulated by the interactions of amino acid sequences within domain 1. *Proc. Natl. Acad. Sci. U. S. A. 99*, 8494-8499.

[95] Eliseev, B., Kryuchkova, P., Alkalaeva, E., and Frolova, L. (2011) A single amino acid change of translation termination factor eRF1 switches between bipotent and omnipotent stop-codon specificity. *Nucleic Acids Res. 39*, 599-608.

[96] Lekomtsev, S., Kolosov, P., Bidou, L., Frolova, L., Rousset, J. P., and Kisselev, L. (2007) Different modes of stop codon restriction by the *Stylonychia* and *Paramecium* eRF1 translation termination factors. *Proc. Natl. Acad. Sci. U. S. A.104*, 10824-10829.

[97] Inagaki, Y., Blouin, C., Doolittle, W. F., and Roger, A. J. (2002) Convergence and constraint in eukaryotic release factor 1 (eRF1) domain 1: the evolution of stop codon specificity, *Nucleic Acids Research 30*, 532-544.

[98] Seit-Nebi, A., Frolova, L., and Kisselev, L. (2002) Conversion of omnipotent translation termination factor eRF1 into ciliate-like UGA-only unipotent eRF1. *EMBO Rep. 3*, 881-886.

[99] Lobanov, A., Turanov, A., Hatfield, D., and Gladyshev, V. (2010) Dual functions of codons in the genetic code. *Critical Rev. Biochem. Mol. Biol. 45*, 257-265.

[100] Skogerson, L., and Wakatama, E. (1976) A ribosome-dependent GTPase from yeast distinct from elongation factor 2. *Proc. Natl. Acad. Sci. U. S. A. 73*, 73-76.

[101] Ypma-Wong, M. F., Fonzi, W. A., and Sypherd, P. S. (1992) Fungus-specific translation elongation factor 3 gene present in *Pneumocystis carinii*. *Infection Immunity 60*, 4140-4145.

[102] Qin, S., Xie, A., Bonato, C. M., and McLaughlin, C. S. (1990) Sequence analysis of the translation elongation factor 3 from *Saccharomyces cerevisiae*. *J. Biol. Chem. 265*, 1903-1912.

[103] Di Domenico, B. J., Lupisella, J., Sandbaken, M., and Chakraburtty, K. (1992) Isolation and sequence analysis of the gene encoding translation elongation factor 3 from *Candida albicans*. *Yeast 8*, 337-352.

[104] Skogerson, L. (1979) Separation and characterization of yeast elongation factors. *Methods Enzymol. 60*, 676-685.

[105] Andersen, C. B. F., Becker, T., Blau, M., Anand, M., Halic, M., Balar, B., Mielke, T., Boesen, T., Pedersen, J. S., Spahn, C. M., Kinzy, T. G., Andersen, G. R., and Beckmann, R. (2006) Structure of eEF3 and the mechanism of transfer RNA release from the E-site. *Nature 443*, 663-668.

[106] Kurata, S., Nielsen, K. H., Mitchell, S. F., Lorsch, J. R., Kaji, A., and Kaji, H. (2010) Ribosome recycling step in yeast cytoplasmic protein synthesis is catalyzed by eEF3 and ATP. *Proc. Natl. Acad. Sci. U. S. A. 107*, 10854-10859.

[107] Gamberi, C., and Lasko, P. (2012) The Bic-C family of developmental translational regulators. *Comp. Funct. Genom.* In press.

[108] Vazquez-Pianzola, P., and Suter, B. (2012) Conservation of the RNA transport machineries and their coupling to translation control across eukaryotes. *Comp. Funct. Genom.* In press.

[109] Layana, C., Ferrero, P., and Rivera-Pomar, R. (2012) Cytoplasmic ribonucleoprotein foci in eukaryotes: hotspots of bio(chemical)diversity. *Comp. Funct. Genom.* In press.

Chapter 3

Evolutionary Molecular Engineering to Efficiently Direct *in vitro* Protein Synthesis

Manish Biyani, Madhu Biyani, Naoto Nemoto and Yuzuru Husimi

Additional information is available at the end of the chapter

http://dx.doi.org/10.5772/52311

1. Introduction

With the completion of human genome project in 2003, the 50th anniversary year of the discovery of the structure of DNA, we entered in the post-genomic era that concentrates on harvesting the fruits hidden in the genomic text. Since then we have witnessed the generation of a tremendous volume of DNA information (*genetic information*). As of September 2011, the Genomes OnLine Database (GOLD, http://www.genomesonline.org) has documented 1914 complete genome projects which comprise 1644 bacterial, 117 archaeal and 153 eukaryal genomes [1]. However, only a fraction of these DNA data are associated with their encoded proteins, i.e., their phenotypes (*functional information*) [2]. Even when a phenotype is associated with the encoding gene, the function of a particular gene cannot be fully understood until it is possible to describe all of the phenotypes that result from the wild-type and mutant forms of that gene. Moreover, unlike a genome that contains a fixed number of genes, the levels of proteins within cells are likely an order of magnitude greater than the number of genes. Therefore, the focus of the scientific community has recently been shifted from gene sequencing to annotation of gene function and regulation through elucidation of protein abundance, expression, post-translational modifications, and protein-protein interactions. While the pre-genomic era which lasted less than 15 years, the post-genomic era can be expected to last much longer, probably extending over several generations, and thus there is an increasing need for high throughput expression of the genome encoded proteins to profile the entire proteome and get a deeper understanding of protein abundance and reveal novel protein functions. Protein synthesis is therefore a powerful tool for large-scale analysis of proteins for a large-variety of low- and high-throughput applications (see Fig.1) and an essential tool for bridging the gap between genomics and proteomics in the post-genomic era. Noteworthy, the ribosome that catalyses and provide the platform for protein synthesis was in the spotlight recently, as the Nobel

Prize in Chemistry 2009 was awarded to the work that unlocked the structure and function of the ribosomes.

2. Significance of protein synthesis enhancer sequences (5'UTR)

Cell-based (*in vivo*) and cell-free (*in vitro*) methods have been developed for production of protein synthesis [3]. Cell-based host systems such as bacteria, yeast, worms, mammalians used for protein synthesis and protein expression analysis, however, have been unable to meet the requirement of producing large amounts of purified and functional proteins which is a prerequisite to facilitate structure-based functional analysis. For example, purified proteins are necessary to grow protein crystals whose X-ray diffraction patterns provide the most precise structural information. Other limitations in host organisms includes such as bacteria don't have the intracellular organelles found in eukaryotes; yeast lack a dimension of complexity in intracellular communication observed in metazoans; and even other mammalian system are different from human in important aspects of both normal physiology and disease pathogenesis [4]. In addition, many biochemical pathways are simply difficult to study in the larger context of other events happening at the same time within the cell. In contrast to the cell-based systems, cell-free protein expression systems are now becoming the favored alternative with far greater fidelity as it offers a simple and flexible system for the rapid synthesis of functional proteins. There is currently a wide range of cell-free translation systems due to the ready availability of cell extracts prepared from various cell sources, including *Escherichia coli*, yeast, wheat germ, rabbit reticulocytes, *Drosophila* embryos, hybridomas, and insect, mammalian, and human cells [5-11]. Although encouraging, there would be some major issues in the use of cell-free systems. First, a major

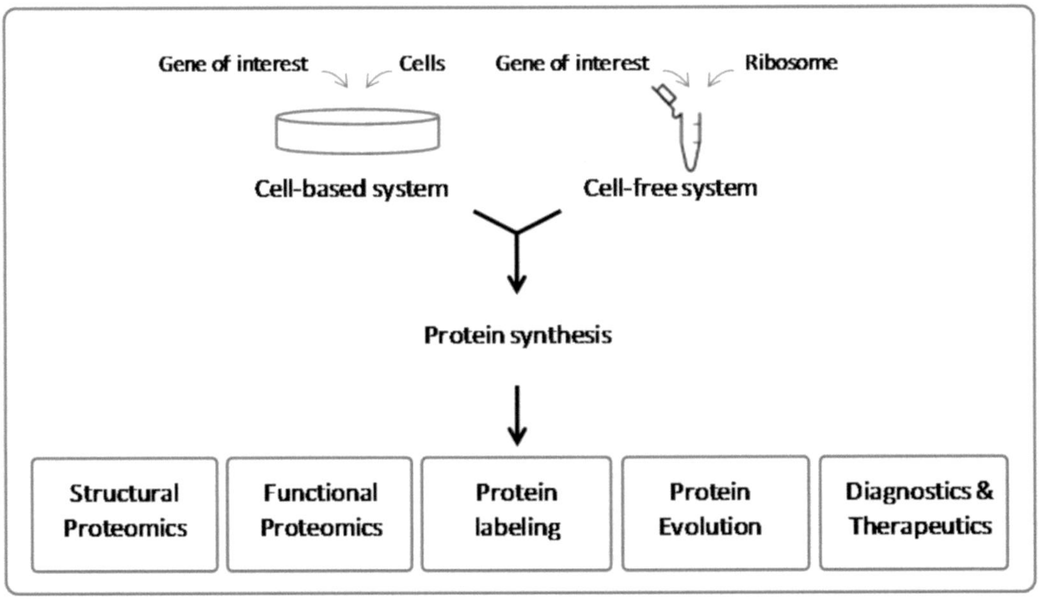

Figure 1. Application of Protein synthesis

drawback of synthesizing proteins in the lysate is that the lysate contains a large portion of the cellular proteins and nucleic acids that are not necessarily involved in the targeted protein synthesis and can lead to low protein yields through interfering with the subsequent purification reactions. In addition, the presence of proteases and nucleases in the lysates could be inhibitory to protein synthesis. In order to addressing this issue, cell-free protein synthesis system was reconstituted *in vitro* from purified components of the *E. coli* translation machinery. This system, termed the "protein synthesis using recombinant elements" (PURE) system, contains all necessary translation factors, purified with high specific activity, and allows efficient protein production [12]. Remarkably, this reconstituted system has been shown to catalyze efficient in vitro protein synthesis by providing a much cleaner background than a lysate-based system [13].

The second issue is that existing cell-free systems differ substantially from each other with respect to their efficiency and scalability to produce proteins and therefore these systems has to be programmed for given exogenous mRNA templates. Although different lysates may contain specific cellular factors that promote protein synthesis, a key factor in ensuring high protein production is the use of strong translational enhancer sequences (untranslated regions, UTRs) in the mRNA templates, which has long been known to enhance protein production up to several hundred-folds [14]. UTRs are known to play crucial role in the post-transcriptional regulation of gene expression, including modulation of the transport of mRNAs out of the nucleus and of translation efficiency [15]. The average length of UTRs motifs located at the 5′end of the exon, called 5′-UTR, ranges between 100 and 200 nucleotides and strikingly varies a lot within a species, e.g., in humans, the longest known 5′UTR is 2,803 while the smallest is just 18 nucleotides [16,17] (Fig.2).

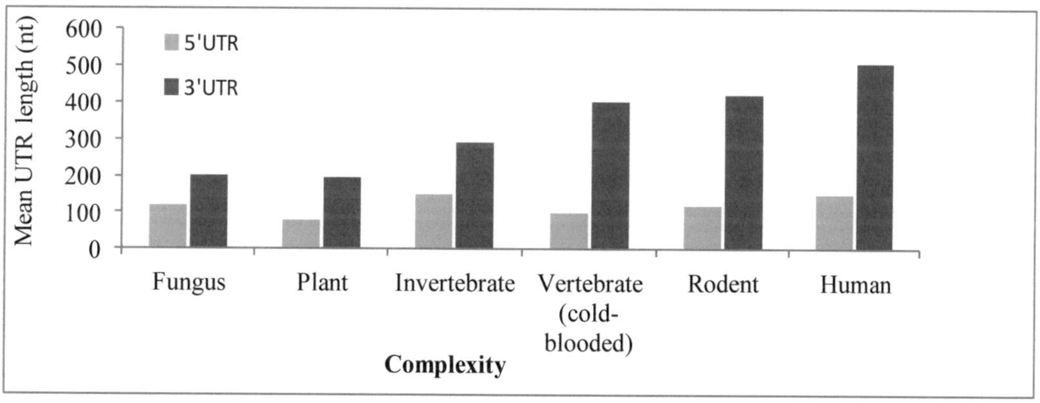

Figure 2. The average length of untranslated region sequences in the different taxonomic classes. Grey bar representing 5′-UTR and black bar is 3′-UTR.

The structural features of the 5′UTR have a major role in the control of protein synthesis. Those proteins which are involved in developmental processes, including growth factors, transcription factors or proto-oncogenes, often have longer 5′UTR than an average and thus untranslated regions of mRNAs have crucial roles in protein regulation through protein synthesis. Structural elements of the eukaryotic mRNA, including the 5′cap and 3′poly(A) tail, and a series of protein-mRNA and protein-protein interactions, including several eIF (eukaryotic initiation factors), are important determinants of translation initiation (Fig.3). In eukaryotes, a multifactor complex of eukaryotic initiation factors are involved in the initiation phase of protein synthesis. But, in particular, 5′UTR plays a major role in the translation initiation, a critical step in protein synthesis which is determining qualitatively and quantitatively which proteins are made, when and where. 5′UTR is composed of several regulatory elements, including the Shine-Dalgarno (SD) and the AU-rich sequences which facilitates 16s rRNA-specific ribosome binding to initiate the protein synthesis [18,19]. In cell-based or in vivo systems, the translation of natural mRNAs is finely regulated by several mechanisms using 5′-capped and 3′-poly(A) containing long-untranslated regions (UTRs).

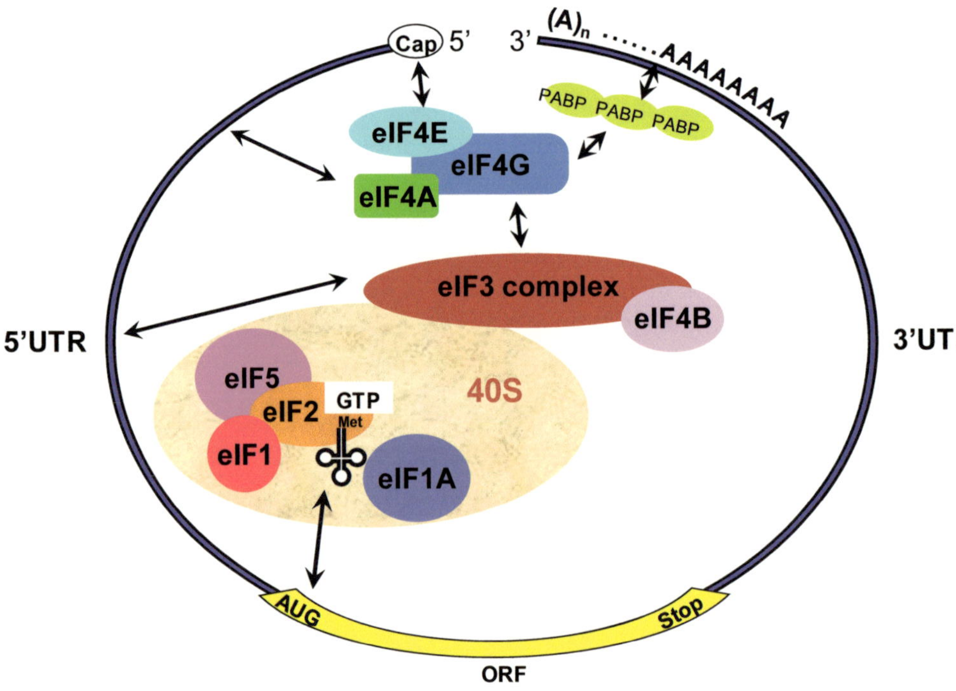

Figure 3. A schematic drawing of a eukaryotic mRNA, illustrating some post-transcriptional regulatory elements that affect initiation of protein synthesis.

Therefore, the efficiency of a cell-free translation system which is reconstituted using crude cell extract is restricted due to the problematic of maintaining long-natural UTRs in the in vitro construct. Even if so, the obvious question here is that "are the natural UTRs can meet the requirements of various translation factors in a cell-free system to carry similar mechanisms as in in vivo system?" Looking at this 'black box' may open a new window into the post-transcriptional regulation of gene expression using cell-free translation systems.

Therefore, it is prerequisite to find an alternate for natural UTRs dependency and optimization of translation initiation in cell-free system for next-generation in vitro high throughput protein synthesis systems. In a recent study using cell-free systems, the translation-enhancing activity of some commonly used natural enhancer sequences, such as omega from tobacco mosaic virus and the 5'UTR of β-globin mRNA from *Xenopus laevis*, was reported to vary from 1- to 10-fold, depending on the source of the cell-free extract used (e.g., wheat germ, rabbit reticulocyte lysate, insect) [20]. Therefore, optimization of enhancer sequences of an exogenous mRNA template with a given crude cell extract is desirable before using a cell-free protein synthesis system. A recent new development has been the remarkable generation of a universal cell-free translation system that mediates efficient translation in multiple prokaryotic and eukaryotic systems by bypassing the need for early translation initiation steps [21].

3. Co-evolutionary relationship between translational initiation and protein synthesis

In the course of evolution on the Earth, how the early life evolved beginning with a hypothetical RNA world-to-the world we know today (DNA world) is the persistent issue of debate for evolutionary biologist. In 1968, Francis Crick argued about the existence of the RNA world in the initial stage of evolution in which RNA molecules assembled from a nucleotide soup and supposed to carry both the genetic and catalytic information (Fig.4). In later stage, some special types of RNA molecules (now termed as Ribozymes) was considered to catalyzes its own self-replication and therefore to develop an entire range of enzymatic activities to form DNA world through an intermediate RNP (RNA/Protein) world. However, there are certain questions that cannot be answered with proposed RNP world. These include: 1. How did 'RNA-world (Ribozyme-type)' evolved to 'DNA-world (cell-type)' since there is no record exists of the intermediates between the RNA-world and organized complexity of cell? 2. What was the first Protein evolved out of an RNA world? 3. How could it have evolved and how the process of translation emerged? 4. If ribosome make protein then how the first ribosomal protein appeared? 5. Why is ribosome made half of protein and half of RNA ?

The recent advances in evolutionary molecular engineering have revealed the bonding strategy of the genotype to its phenotype as a unique and essential nature of a 'virus' and thus the role of virus-type strategy in the course of evolution on the Earth. In 1995, Nemoto and Husimi proposed a 'virus-early and cell-late model' that a virus-like molecule consist of genotype (mRNA) and phenotype (its coded protein) molecules emerged in the latter period of RNA world was the key molecule which enforced the transition from RNA-to-RNP world by co-evolving the translation system and a virus-like molecule coded a primitive protein of replicase [22]. In this theory, they also showed that such virus-like molecule could introduce Darwinian evolution into the Eigen's hypercycle members (RNA replicase of RNA, RNA translation members, RNA replicase of protein) resulting in carrying out co-evolution between translation system and protein replicase. This was later reinforced by inventing and demonstrating a genotype-phenotype linked method (IVV, in vitro virus) for evolutionary molecular engineering [26] and this strongly suggest the potential of IVV method to understand the relation between ribosome-mRNA interaction.

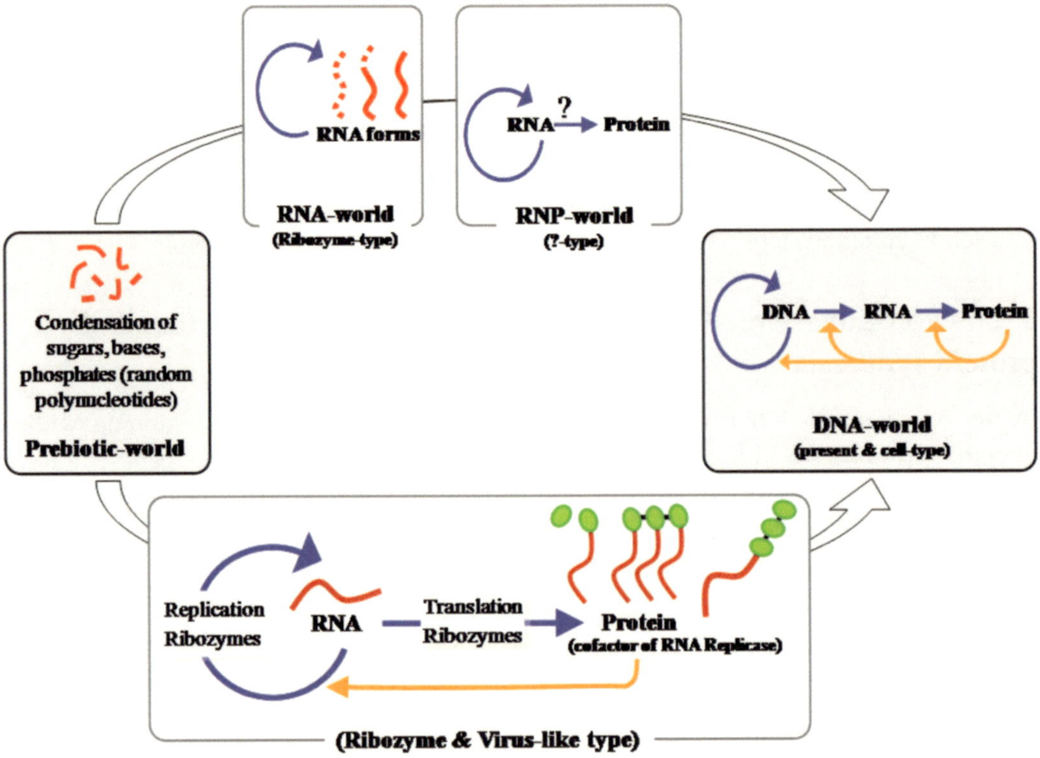

Figure 4. A schematic drawing to represent co-evolution of the translation initiation and protein synthesis system, prior to 'birth of first cellular life'.

4. Directed molecular evolution and screening of protein synthesis enhancer sequences

Directed molecular evolution mimics the natural Darwinian evolution process to evolve new functional molecules in the laboratory rather than in the jungle and in days rather than in millenniums and thus has emerged as a dominant approach for exploiting the sequence space to generate biomolecules with novel functions. Directed molecular evolution rely on the application of selection pressure to identify a bio-molecule with desirable properties from a diverse pools (or 'libraries') of bio-molecules with hundreds of millions of mutations and consist of four essential and repeating cycles: the creation of mutation and diversity at the DNA molecular level; the coupling of genetic information (DNA/mRNA) to functional information (Protein); the application of selection pressure; and the amplification of selected molecules (Fig.5).

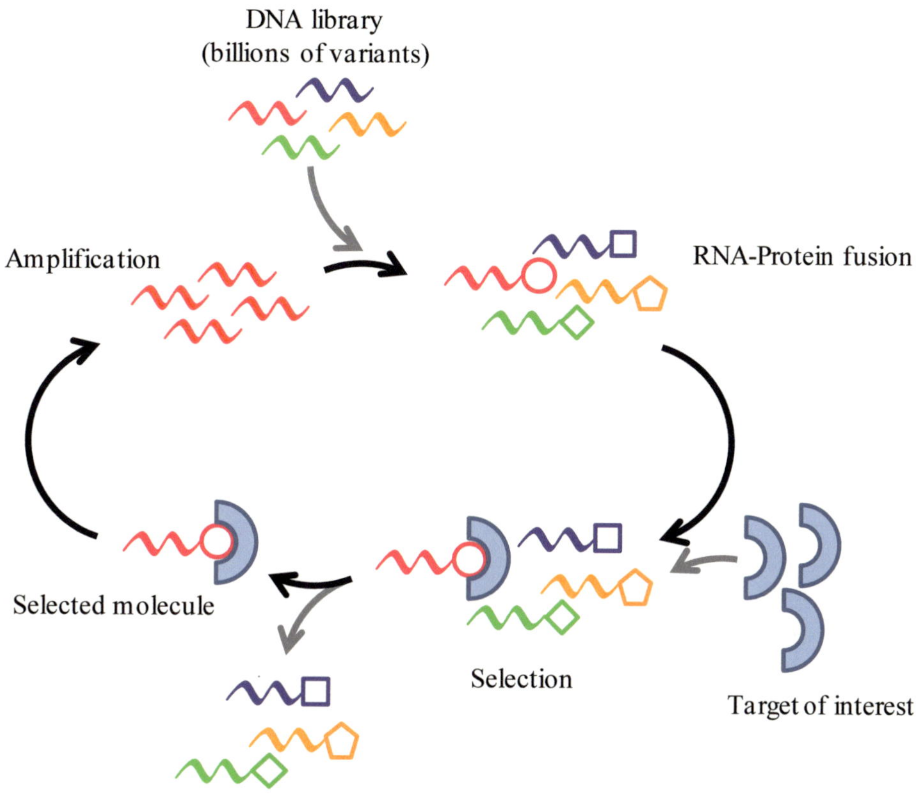

Figure 5. A schematic drawing of Directed Molecular Evolution

A number of well-established strategies, called display technologies, have been developed which use natural cell-based environment, such as yeast surface display, bacterial surface display, phage display or use a cell-free environment, such as ribosome display, mRNA display (in vitro virus), cDNA display, CIS display, IVC (in vitro compartmentalization) (Fig.6).

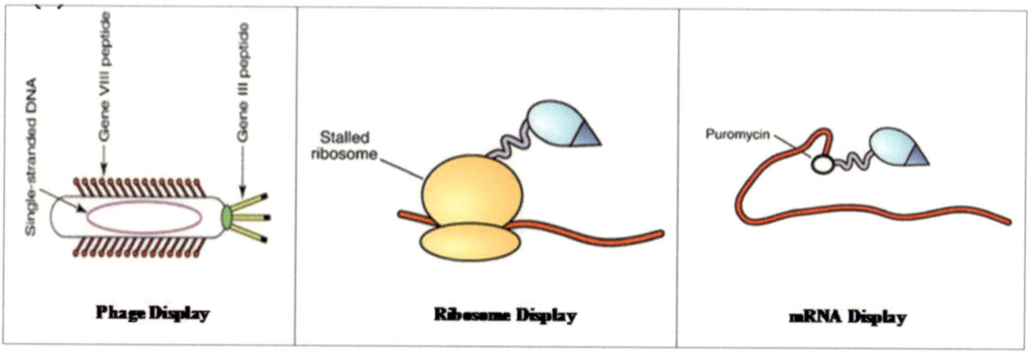

Figure 6. A schematic drawing of well-known strategies to perform Directed Molecular Evolution in the laboratory

Interestingly, a few groups have reported the application of directed molecular evolution to the screening of enhancer sequences with high translation efficiency in a cell-free translation system using ribosome display or polysome-mediated selection methods [23-25]. Recently, a novel strategy is also described for the in vitro selection of strong translation enhancer sequences for use in a cell-free translation system using an mRNA display method. The mRNA display method (originally called an "in vitro virus") [26,27], which covalently links the mRNA molecule (genotype) to its encoded protein (phenotype), is a powerful evolutionary method for searching for functional protein molecules in a large-scale library. In this strategy, a simplified new gel shift assay system was developed to demonstrate that short but efficient translation enhancer sequences can be created for use in a given cell-free translation system (Fig.7). This method is based on an mRNA display method in which a covalent linkage is formed between the mRNA and the encoded protein through the antibiotic molecule puromycin. The steps involved in the synthesis of the covalently linked mRNA–protein fusion, and in the selection of 5′UTR sequences, are summarized below. First, a model gene construct is designed (Fig.7A) as a positive control (wt), which consists of a T7 promoter and a natural 5′UTR sequence (*X. laevis b-globin*) upstream of the PDO coding sequence. The stop codon is deleted to facilitate RNA–protein fusion, and a short DNA fragment complementary to a Puro-linker DNA sequence is ligated downstream of the coding sequence. Second, a random variable 5′UTR library is constructed by replacing the cognate secondary structure part of the *X. laevis b-globin* UTR sequence (36 nt) with a randomized 20-nt-long sequence with all possible combinations of the four nucleotides (N_{20}) (Fig.7B), resulting in an initial library size of approximately 10^{12} (4^{20}) molecules. Third, the cDNA library is then transcribed into an mRNA library using T7 RNA polymerase with/without the cap analogue (m7GpppG). Fourth, the 3′-terminal end of the mRNA library is ligated to a synthetic Puro-linker DNA. Fifth, the resulting mRNA–Puro-linker conjugate library is then used as a template in a given cell-free translation system and is converted into an mRNA–protein fusion library. Sixth, to select efficient 5′UTR candidates from inefficient ones, the resulting mRNA–protein fusion is analyzed using SDS–PAGE. As shown in Fig.7F, fusion products (translated products) of efficiently translated 5′UTRs will migrate with a decreased mobility compared with untranslated products from 5′UTR regions with no and/or slow translation efficiency. Thus, translated and nontranslated candidates can be distinguished, and translated candidates can be clearly identified, by a shift in the gel band pattern. Seventh, the fusion product of translated candidates is then carefully excised from the gel, and the associated mRNAs that represent selected 5′UTR candidates for efficient translation are directly reverse-transcribed and amplified using a single-step RT–PCR. This PCR step completes one round of selection. Finally, the selected 5′UTR candidates are then used as templates for a subsequent selection round for further enrichment of efficient 5′UTR sequences. Using this gel-shift assay, the translation of an mRNA template using a population of randomized 20-nt-long sequences upstream of a Pou-specific DNA-binding domain of Oct-1 (PDO) was screened with a rabbit reticulocyte extract and the time for translation was successively shortened. A total of five selection rounds were performed, starting with a translation time of 45 min and reducing the time by 10 min for each subsequent round. The final round used a translation time of only 5 min. The total yield of RNA–protein fusion constructs following translation after each round was evaluated using SDS–PAGE analysis and reported to gradually increased with each successive round of selection [28].

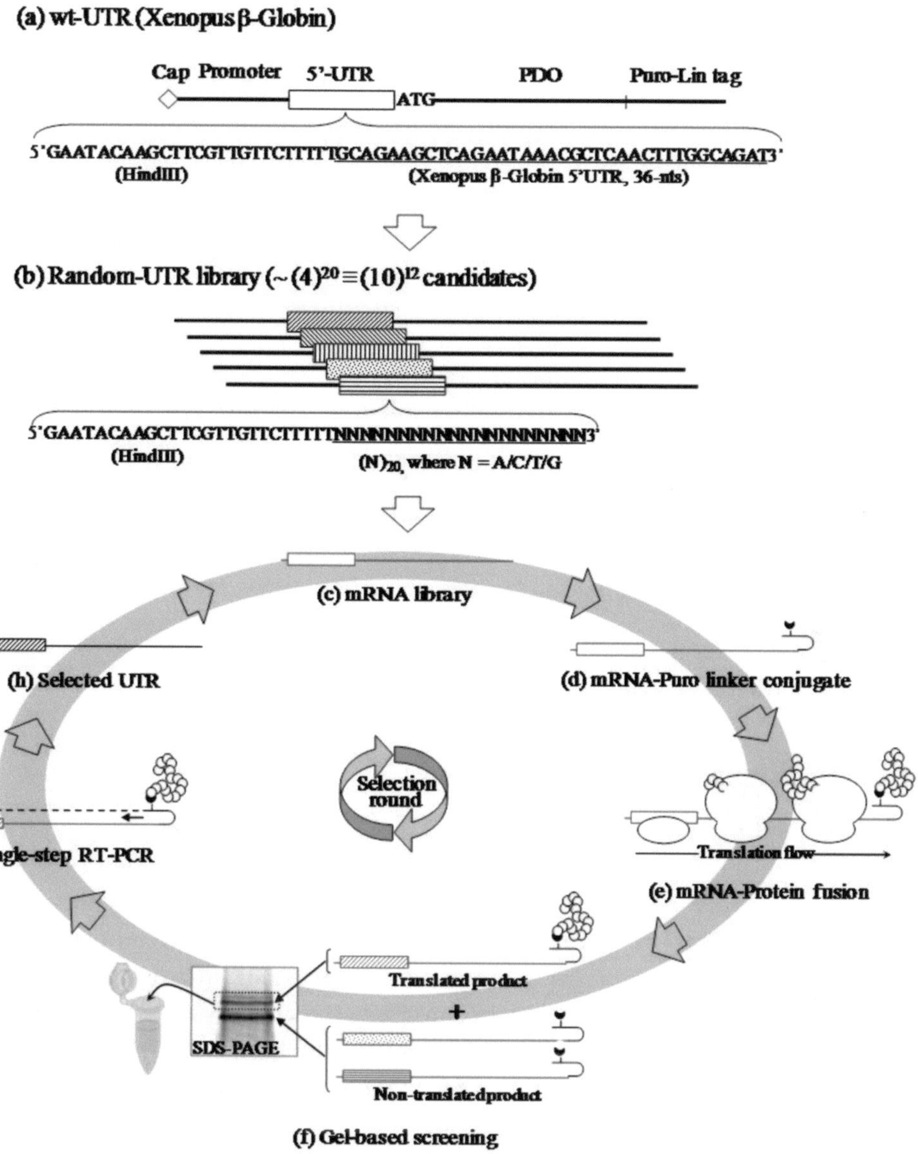

Figure 7. Schematic representation and flow diagram of a novel gel-shift selection method for searching a strong translation enhancer sequence against a given cell-free translation system using mRNA display. DNA template constructs used in the screening experiments including known 5′-untranslated region of Xenopus-Globin (wt-UTR) (a) and random-UTR library (b). The mRNA library which is lacking a stop codon (c) is ligated at the 3′-terminus end to the complementary portion of 5′-terminus end of the puromycin-linker DNA (d) and translated in a cell-free translation system (e). The ribosome stalls at the mRNA and linker-DNA junction during translation. This permits puromycin to enter the ribosomal A-site and to bind to the nascent polypeptide chain. Translated products are analyzed by SDS-PAGE analysis and carefully excised from gel to separate them from non-translated products (f). The associated mRNAs which represent the selected 5′UTR candidates were then directly revere-transcribed and amplified using single-step RT-PCR (g) and used as template for next selection-round (h).

This increase confirmed that the selected library is successively enriched for strong translation enhancer sequences after each round of selection and thus the gel shift selection method using mRNA display is indeed a simple and effective method of screening for strong translation enhancer sequences. The analysis of selected sequences showed the richness of T and G bases with an average of 53% and 35%, respectively, indicating a significant role of U and G bases in the translation enhancer sequences. In addition, these selected sequences was confirmed to show higher translation efficiency in comparison with the natural and longer enhancer sequences. These results encouraged that the described gel-shift method could be applied to a rapid screening of novel 5′UTR which can facilitate cap-independent (IRES-mediated) protein synthesis in cell-free translation systems without the assistance of the full set of initiation factors. Very recently, a few interesting 5′UTRs have been proposed to accelerate the translation initiation reaction [29,30]. These findings of simple and effective 5′UTR suggest the possibility of improvement of 5′UTR under the conditions in various cell-free translation systems. Our approach can be applied to the further searching for 5′UTR by combining with these researches. In conclusion, gel-shift method demonstrated that shorter but strong translation enhancer sequences which should be easier to handle than long natural sequences can be selected rapidly by simple and robust mRNA display method. Searching for novel 5′UTR will contribute much toward the development of proteomics and evolutionary protein engineering research by improvements of cell-free translation methodologies.

5. Conclusion and future perspective

This chapter represents a simple, rapid, easy, and novel strategy, called 'Gel-shift selection', to obtain strong translation enhancer sequence variants for tunable protein synthesis using cell-free system. This method can further explore for (i) discovering of nuclease-resistant stable hairpin secondary structure to stabilize the 5′-terminus end of mRNA template with improved half-life instead of using synthetic 5′-cap analog; (ii) optimization of strong translational enhancer motifs which is free of 5′-cap dependency of translation initiation to improve the translational efficiency on given mRNAs under given translational conditions in cell-free system; (iii) optimization of enhancer motifs which is free of 3′-poly(A) dependency to eliminate the poly(A) leader effect which provide the abolition of the inhibition of translation at excess mRNA concentration.

Author details

Manish Biyani[1,2,*], Madhu Biyani[1,3], Naoto Nemoto[3] and Yuzuru Husimi[4]
[1]Department of Biotechnology, Biyani Group of Colleges, R-4, Sector No 3, Jaipur, India
[2]Department of Bioengineering, The University of Tokyo, Tokyo, Japan
[3]Department of Functional Materials Science, Saitama University, Saitama, Japan
[4]Innovative Research Organization, Saitama University, Saitama, Japan

[*] Corresponding Author

6. References

[1] Pagani I, Liolios K, Jansson J, Chen IM, Smirnova T, Nosrat B, Markowitz VM, Kyrpides NC., The Genomes OnLine Database (GOLD) v.4: status of genomic and metagenomic projects and their associated metadata. *Nucleic Acids Res.* 2012, 40 (Database issue):D571-9.

[2] Prasun, P., Pradhan, M., Agarwal, S., One gene, many phenotypes. *J. Postgrad. Med.*, 2007, 53, 257-261.

[3] Yokoyama S., Protein expression systems for structural genomics and proteomics. *Curr Opin Chem Biol.* 2003, 7, 39-43.

[4] Winslow J, Insel T. The social deficits of the oxytocin knockout mouse. *Neuropeptides* 2002, 36, 221 – 229.

[5] Kigawa, T., Yabuki, T., Matsuda, N., Matsuda, T., Nakajima, R., Tanaka, A., Yokoyama, S., Preparation of Escherichia coli cell extract for highly productive cell-free protein expression, *J. Struct. Funct. Genomics* 2004, 5, 63–68.

[6] Hartley A.D., Santos M.A., Colthurst D.R., Tuite M.F., Preparation and use of yeast cell-free translation lysate, *Methods Mol. Biol.* 1996, 53, 249–257.

[7] Erickson A.H., Lobel G., Cell-free translation of messenger RNA in a wheat germ system, *Methods Enzymol.* 1983, 96, 38–50.

[8] Jackson R.J., Hunt T., Preparation and use of nuclease-treated rabbit reticulocyte lysates for the translation of eukaryotic messenger RNA, *Methods Enzymol.* 1983, 96, 50–74.

[9] Mikami S., Kobayashi T., Yokoyama S., Imataka H., A hybridoma-based in vitro translation system that efficiently synthesizes glycoproteins, *J. Biotechnol.* 2006, 127, 65–78.

[10] Ezure, T., Suzuki, T., Higashide, S., Shintani, E., Endo, K., Kobayashi, S., Shikata, M., Ito, M., Tanimizu, K., Nishimura, O., Cell-free protein synthesis system prepared from insect cells by freeze–thawing, *Biotechnol. Prog.* 2006, 22, 1570–1577.

[11] S. Mikami, T. Kobayashi, M. Masutani, S. Yokoyama, H. Imataka, A human cell derived in vitro coupled transcription/translation system optimized for production of recombinant proteins, *Protein Exp. Purif.* 2008, 62, 190–198.

[12] Shimizu Y, Inoue A, Tomari Y, Suzuki T, Yokogawa T, Nishikawa K, Ueda T., Cell-free translation reconstituted with purified components. *Nat Biotechnol.* 2001, 19, 751-755.

[13] Shimizu Y, Kanamori T, Ueda T. Protein synthesis by pure translation systems. *Methods* 2005, 36, 299-304.

[14] Falcone, D., Andrews, D.W., Both the 5′untranslated region and the sequences surrounding the start site contribute to efficient initiation of translation in vitro, *Mol. Cell. Biol.* 1991, 11, 2656–2664.

[15] van der Velden AW, Thomas AA: The role of the 5′untranslated region of an mRNA in translation regulation during development. *Int J Biochem Cell Biol* 1999, 31, 87-106.

[16] Mignone F, Gissi C, Liuni S, Pesole G: Untranslated regions of mRNAs. *Genome Biology* 2002, 3, 1-10.

[17] Pesole G, Mignone F, Gissi C, Grillo G, Licciulli F, Liuni S: Structural and functional features of eukaryotic mRNA untranslated regions. *Gene* 2001, 276, 73-81.

[18] Grunberg-Manago, M., Messenger RNA stability and its role in control of gene expression in bacteria and phages. *Annu. Rev. Genet.* 1999, 33, 193–227.
[19] Salvador, M.L., Klein, U., Bogorad, L., 5′ sequences are important positive and negative determinants of the longevity of Chlamydomonas chloroplast gene transcripts. *Proc. Natl. Acad. Sci. USA* 1993, 90, 1556–1560.
[20] Suzuki, T., Ito, M., Ezure, T., Kobayashi, S., Shikata, M., Tanimizu, K., Nishimura, O., Performance of expression vector, pTD1, in insect cell-free translation system. *J Biosci Bioeng.* 2006, 102, 69-71.
[21] Mureev, S., Kovtun, O., Nguyen, U.T. and Alexandrov, K. Species-independent translational leaders facilitate cell-free expression. *Nat Biotechnol.* 2009, 27, 747-752.
[22] Nemoto, N. and Husimi, Y., A Model of the Virus-type Strategy in the Early Stage of Encoded Molecular Evolution., *J. Theor. Biol.* 1995, 176, 67-77.
[23] Kamura, N. and Sawasaki, T. Selection of 5″UTR that enhance initiation of translation in a cell free protein synthesis system from wheat embryos. *Bioorg Med Chem Lett.* 2005, 15, 5402-5406.
[24] Nagao, I. and Obokata, J. In vitro selection of translational regulatory elements. *Anal Biochem.* 2006, 354, 1-7.
[25] Mie, M., Shimizu, S., Takahashi, F. and Kobatake, E. Selection of mRNA 5'-untranslated region sequence with high translation efficiency through ribosome display. *Biochem Biophys Res Commun.* 2008, 373, 48-52.
[26] Nemoto, N., Miyamoto-Sato, E., Husimi, Y. and Yanagawa, H. In vitro virus: bonding of mRNA bearing puromycin at the 30-terminal end to the C-terminal end of its encoded protein on the ribosome in vitro. *FEBS Lett.* 1997, 414, 405–408.
[27] Roberts, R. W., and Szostak, J. W. RNA-peptide fusions for the in vitro selection of peptides and proteins. *Proc Natl. Acad. Sci. USA* 1997, 94, 12297-12302.
[28] Biyani, M., Biyani, M. Nemoto, N., Ichiki, T., Nishigaki K., Husimi, Y. Gel-shift selection of translation enhancer sequences using mRNA display. *Anal Biochem* 2011, 409, 105-11.
[29] Shirokikh, N.E. and Spirin, A.S. Poly(A) leader of eukaryotic mRNA bypasses the dependence of translation on initiation factors. *Proc Natl Acad Sci U S A.* 2008, 105, 10738-10743.
[30] Elfakess, R. and Dikstein, R., A translation initiation element specific to mRNAs with very short 5'UTR that also regulates transcription., *PLoS One* 2008, 3, e3094.

Section 3

Cell-Free System and Protein Synthesis

Chapter 4

Protein Synthesis *in vitro*: Cell-Free Systems Derived from Human Cells

Kodai Machida, Mamiko Masutan and Hiroaki Imataka

Additional information is available at the end of the chapter

http://dx.doi.org/10.5772/48563

1. Introduction

When researchers wish to obtain recombinant proteins, a primary choice of the method is in most cases the expression in *E. coli*. If this system does not work for the protein of interest, they may turn to insect or mammalian cells. Protein expression *in vitro* may be chosen if these *in vivo* expression systems do not give the protein satisfactorily. There are several reasons why expression of some recombinant proteins in living cells is poor. If the protein to be expressed is toxic to host cells or inhibitory for growth, it should be difficult to express the protein to a high level. The cell-free system is derived from the extract from broken cells, and therefore the above-mentioned problems that occur in the living cells, if not all, are avoidable.

The value of the cell-free protein system seems unlimited. Radioactive amino acids or unnatural amino acids are relatively easily incorporated into proteins, rendering the system very useful for structural analysis of the synthesized proteins (1). The cell-free translation systems have also been used in the high-throughput production of thousands of gene products derived from cDNA libraries to facilitate screening in the identification of kinase or proteinase targets. While *E. coli*- (2), wheat germ- (3) and rabbit reticulocyte- (4) derived systems have been widely employed for the above-mentioned purposes for decades, human cells-derived *in vitro*-protein expression systems are now beginning to gain attention.

What is the merit of a human or mammalian cells-derived system compared with other cell-free systems? Firstly, many different cell lines that are derived from various organs or tissues such as neurons, endocrine glands and immunocytes are available from cell banks (eg., ATCC and RIKEN BRC). Since each cell line maintains some properties specific to the originated organs or tissues, one can establish a variety of cell-free systems from different cell lines. A successful example is a cell-free glycoprotein-synthesis system derived from a monoclonal antibody-producing hybridoma (5). Another merit of the mammalian system is that mammalian cell-derived extracts seem to have greater capacity to synthesize large proteins (6) than other systems. Lastly, mammalian cell-free systems can directly lead to application for medical and pharmaceutical purposes. A remarkable example is the synthesis of RNA virus in a test tube (7), which is impossible by other cell-free systems. The RNA virus is a super-high molecular weight complex consisting of its RNA genome and capsid proteins, and the viral particles are assembled through a complex process. The assembly process of the RNA virus such as picornaviruses can be recapitulated in a human cell-derived *in vitro* protein synthesis system. This system can be used for screening anti-virus chemicals. In the following sections, we discuss these three advantages of the human cell-based in vitro protein synthesis systems.

2. Cell-free synthesis of glycoproteins

Glycosylation is one of the major post-translation modifications of proteins. The polypeptides destined to be localized to the plasma membrane or to be secreted outside of the cell enter the endoplasmic reticulum (ER) while being translated. Immediately after the polypeptide enters the ER, *N*-linked glycosylation starts (8). Whereas glycans linked to proteins are implicated in many biological aspects such as development, differentiation and physiology (9), *N*-glycosylation itself is thought to be necessary for the proper folding of proteins through disulfide bond formation in the ER (8). More than half of eukaryotic proteins are predicted to be glycoproteins(10), and thus, it is an urgent task to establish an efficient system to produce glycoproteins *in vitro*. However, neither the *E. coli* nor wheat germ system can glycosylate proteins. rabbit reticulocyte lysates (RRL) combined with microsomes from dog pancreas have been a popular system for *in vitro N*-glycosylation (11), but commercially available RRL and canine microsomes are expensive, and the activities of the preparations vary depending on the source. Furthermore, preparation of these extracts by a researcher is not an easy task, since these systems entail sacrifice of animals.

These problems can be solved by using a specific cell line. HeLa cells represent one of the most popular cell lines as a source of mammalian cell-derived *in vitro* translation systems. However, HeLa cell extracts (12) fail to produce a recombinant glycoprotein (5). The major envelope glycoprotein (gp120) of human immunodeficiency virus type-1 (HIV-1) consists of a core polypeptide of ~60 kDa and ~20 *N*-linked glycans which increase the total mass of the molecule to ~120 kDa (13). When mRNA encoding the HIV-1-gp120 region is translated in the HeLa cell extract, a ~60 kDa protein is synthesized as the major product. This indicates that *N*-glycosylation of gp120 is inefficient in the HeLa cell-derived cell-free system. The endoplasmic reticulum (ER), where *N*-glycosylation occurs, may not be well-developed in HeLa cells, because HeLa is not a secretory cell line. A monoclonal antibody-producing hybridoma cell line is now chosen, because the hybridoma cell should possess a highly developed ER system to secrete large amounts of immunoglobulins, and, from the practical point of view, they can be easily propagated in a suspension culture (5). When programmed with the mRNA encoding HIV-1-gp120, the hybridoma extract prominently synthesizes one product with a molecular mass of ~100 kDa (Figure 1). This 100 kDa product is a glycosylated form of gp120, since treatment with PNGase F changed it into a ~60 kDa product (Figure 1). Other than HIV-1-gp120, biologically active human choriogonadotropin (hCG), a glycoprotein hormone consisting of α and β subunits was successfully synthesized (5).

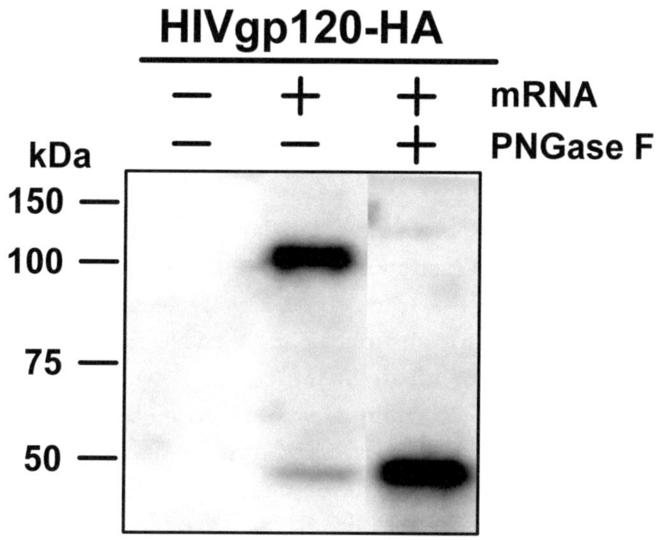

Figure 1. Glycosylation of HIV gp120 in the hybridoma-derived cell-free translation system.
The hybridoma extract was incubated with mRNA encoding gp120-HA. After incubation, samples were treated with or without PNGase F, resolved by SDS-PAGE, and analyzed by western blotting with anti-HA antibody.

3. Cell-free synthesis of large proteins

Many human proteins are very large (>150 kDa), and these large proteins consist of several functional domains. Each domain may be expressed by conventional protein expression systems such as in *E. coli* for the functional and structural analysis. However, it is obviously necessary to examine the structure and function of the whole molecule to gain insight into the real function of the protein, yet many large proteins remain uncovered for the structure and function due to the difficulty in preparation of the full-length form. Since the mammalian cells have many large proteins, it is expected that the mammalian translation machinery basically possesses a high capacity to elongate a long chain of the peptide. Translation initiation is the limiting step in the eukaryotic protein synthesis, and therefore the capacity to elongate a peptide chain of thousands amino acids can be recapitulated *in vitro* if the initiation step is not impaired.

Among the factors involved in translation initiation, eukaryotic translation initiation factor (eIF) 2 plays a pivotal role in translational regulation (14). eIF2 comprises three subunits: α, β and γ. A ternary complex consisting of eIF2-GTP-methionyl initiator tRNA (Met-tRNAi^{Met}) transfers Met-tRNAi^{Met} to the 40S ribosomal subunit. When the anticodon of Met-tRNAi^{Met} base-pairs with the AUG initiation codon, the eIF2-bound GTP is hydrolyzed to GDP, and eIF2-GDP is subsequently released from the ribosomal complex. For the next round of translation initiation, eIF2-GDP must be converted to eIF2-GTP to regenerate the ternary complex, a reaction catalyzed by eIF2B, a multi-protein complex with 5 subunits. When the α subunit of eIF2 is phosphorylated, the affinity of eIF2 for eIF2B dramatically increases, and eIF2B is thereby sequestered by eIF2. Since eIF2B is then unable to regenerate the ternary complex, translation is consequently attenuated (14). Phosphorylation of the α subunit of eIF2 occurs in response to stress conditions such as viral infection, oxidation, deprivation of amino acids, and accumulation of misfolded proteins (15).

The mammalian cell extract-derived *in vitro* translation system has an intrinsic problem, namely, phosphorylation of the α subunit of eIF2 due to a high concentration of ATP (5). During preparation of the cell extract, eIF2α-kinases seem to aggregate or dimerize, and addition of ATP causes auto--phosphorylation and activation of the kinases. The ATP-induced phosphorylation of eIF2α in the cell-free system is a serious problem, because phosphorylation of eIF2α attenuates translation initiation (14), yet, ATP is indispensable to maintain protein synthesis. This problem is now solved by addition of K3L and GADD (growth arrest and DNA damage gene) 34. K3L is a vaccinia virus-encoded protein, which acts as a pseudosubstrate for eIF2α-kinases because of its structural resemblance to an N-terminal part of eIF2α (16, 17). K3L prevents phosphorylation of eIF2α during the virus infection, thereby counteracting the otherwise repressed translation (18). GADD34 recruits a phosphatase PP1c to eIF2 to dephosphorylate eIF2α (19). As expected, addition of recombinant K3L and / or GADD34 relieves phosphorylation of eIF2α and effectively stimulates protein synthesis in the cell-free system (5, 20).

The cell-free protein expression system supplemented with K3K/GADD34 is further improved by introduction of coupled transcription/translation system and internal ribosome entry site (IRES). In cell-free translation systems, mRNA is added or it is synthesized with the addition of a DNA (a plasmid or a PCR product) and the bacteriophage RNA polymerase (T7, SP6, or T3 RNA polymerase). The latter method, called a coupled transcription/translation system (6), is more convenient than the mRNA-dependent system, because researchers do not need to synthesize and purify RNA. Furthermore, mRNA is continuously supplied to compensate for degradation of the mRNA in the system.

A drawback of the coupled transcription/translation system is that RNA produced by a bacteriophage RNA polymerase is not 5′-capped unless a high concentration of the cap-analogue is supplied. Uncapped RNAs are less efficient for translation than the capped counterpart in the HeLa cell-derived cell-free system, and the cap-analogue is very expensive. This problem is solved by placing the encephalomyocarditis virus (EMCV) internal ribosome entry site (IRES) or the hepatitis C virus (HCV) IRES between the T7 promoter and the coding region of the plasmid. The ribosome binds to IRES and initiates translation without aid of the cap structure.

Collectively, the HeLa cell-based *in vitro* coupled transcription/translation system supplemented with K3L/GADD34 is able to synthesize large proteins such as GCN2 (170 kDa), Dicer (200 kD), eIF4G (220 kDa) and mTOR (260 kD) from the IRES-harboring plasmids that encode for these proteins (Figure 2).

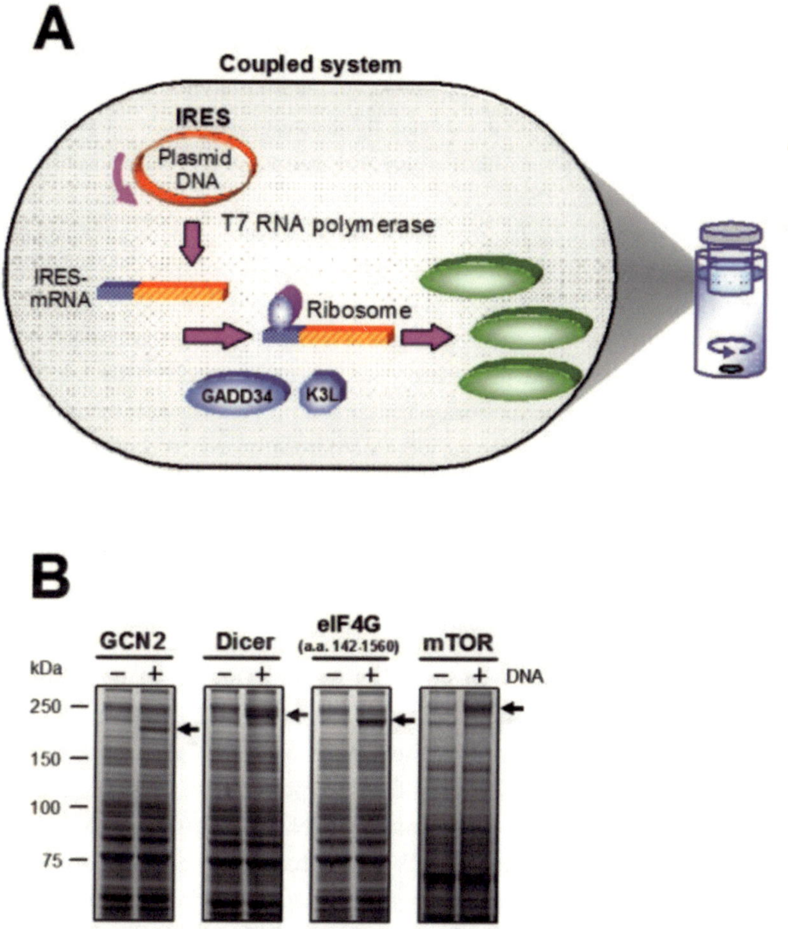

Figure 2. The HeLa cell-based in vitro coupled transcription/translation system.
(A) Cartoon depicting the HeLa cell-based in vitro coupled transcription/translation system that utilizes IRES and is supplemented with K3L and GADD34. (B) Large proteins were synthesized using the system depicted in (A). Samples were separated by SDS-PAGE and stained with CBB. Arrows indicate synthesized proteins.

4. Cell-free synthesis of RNA virus

Cell-free synthesis of an infectious virus is an ideal tool for elucidating the mechanism of viral replication and for screening anti-viral drugs. *Encephalomyocarditis virus* (EMCV) is a *Cardiovirus* in the family *Picornaviridae*. The genome of EMCV is a single-stranded positive-sense RNA of 7.9 kb. Upon infection by EMCV, the genomic RNA is translated into a single polyprotein, which is subsequently processed via a series of proteolytic events into structural (capsid) and nonstructural proteins (21). The RNA-dependent RNA polymerase (RdRp), one of the viral nonstructural proteins, replicates the genomic RNA, which is incorporated into a viral capsid intermediate structure to constitute an infectious virion (22) (23). EMCV can be synthesized from its RNA in the HeLa extract-derived cell-free protein synthesis systems (24, 25) (Figure 3). Since the synthesized virus is proved to be infectious (24) (Figure 4), the *in vitro* system is a choice of the method to obtain virus particles.

The cell-free synthesis of EMCV is enhanced by employing a dialysis system (Figure 3). A batch system (a closed test tube system) does not allow for sustained synthesis of proteins over a period of several hours due to amino acid and ATP deficiencies, and to the accumulation of waste products. In contrast, a dialysis system, which continuously supplies the substrates and energy source for protein synthesis and removes waste products through a dialysis membrane, has enabled HeLa cell extracts to maintain protein synthesis for up to one day. To investigate the means by which the dialysis system enhances virus synthesis, the efficiencies of translation and processing steps were monitored by labeling with radio-labeled leucine during a 10-h incubation. Neither the processing pattern of the viral proteins nor the intensity of each product varied substantially when HeLa cell extracts were incubated with the viral RNA by the batch or dialysis system (24). In contrast, the capacity of the HeLa cell extract to synthesize EMCV RNA was increased seven-fold by employing the dialysis system compared with the batch system (24). Thus, replication of the RNA, rather than translation or processing of the viral proteins, is enhanced by the dialysis system (Figure 5).

A ribozyme technology provides opportunities for mutational analyses of EMCV *in vitro* and for production of EMCV particles (24, 26) (Figure 3). Efficient RNA synthesis by the virus-encoded RdRp requires a precise sequence at the 5′-end of the template RNA. Synthetic RNAs produced by T7, SP6, and T3 RNA polymerases have a guanine nucleotide at the 5′-end, which hampers the plus stranded RNA synthesis of EMCV. No detectable virus was generated from a synthetic EMCV RNA that possessed extra nucleotides (GG) at the 5′-end (24). If genomic RNA purified from EMCV particles were the only available template, the usefulness of the cell-free system would be limited, because mutational studies could not be done. Thus, a hammerhead ribozyme sequence was introduced at the 5′-end of the RNA to catalyze removal of the extra nucleotides; introduction of an appropriately designed hammerhead ribozyme sequence at the 5′-end of the RNA yields an RNA with the same nucleotide sequence at the 5′-end as the viral genomic RNA (24). Synthetic EMCV

RNAs were translated with comparable efficiencies in HeLa cell extracts by the dialysis method irrespective of the presence of the ribozyme at the 5′ end. However, while replication of synthetic EMCV RNA without the ribozyme was not appreciable, the EMCV RNA with the ribozyme replicated at 25% the efficiency of the genomic EMCV RNA (24) (Figure 6).

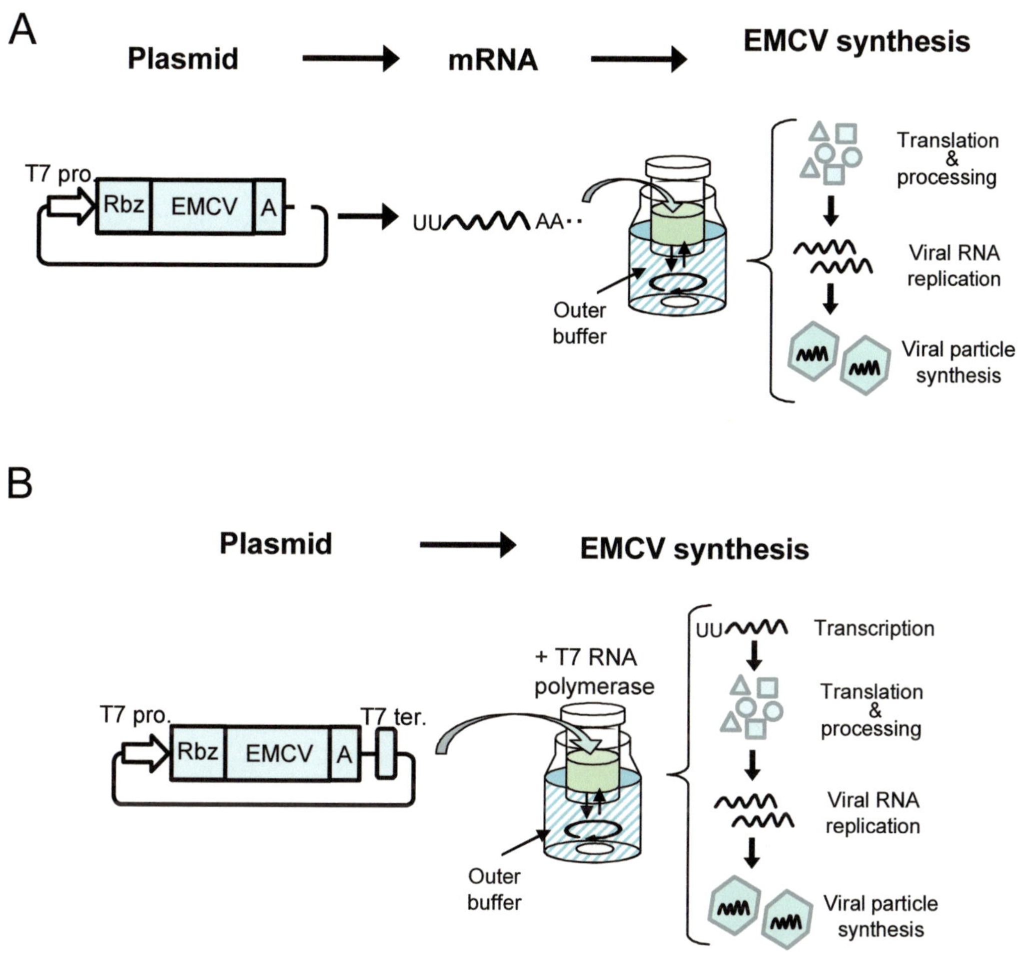

Figure 3. RNA-dependent and DNA-dependent cell-free systems for EMCV synthesis
(A) mRNA dependent system. EMCV RNA is synthesized in vitro and purified. The purified EMCV RNA is incubated with the HeLa cell-derived cell-free protein synthesis system. (B) DNA-dependent system. The plasmid encoding the EMCV RNA is directly incubated with the HeLa cell-derived cell-free protein synthesis system supplemented with T7 RNA polymerase.

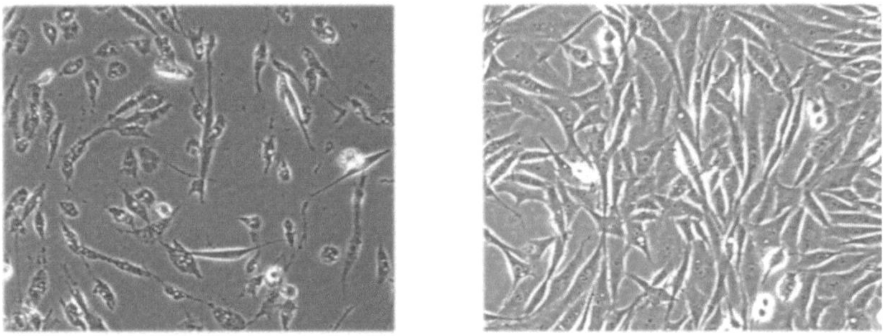

Figure 4. Infection of BHK-21 cells with EMCV synthesized by the cell-free system.
BHK-21 cells were incubated with RNase-treated HeLa cell extract programmed with (left panel) or without (right panel) EMCV RNA. Twenty hours later, cells were observed by microscopy.

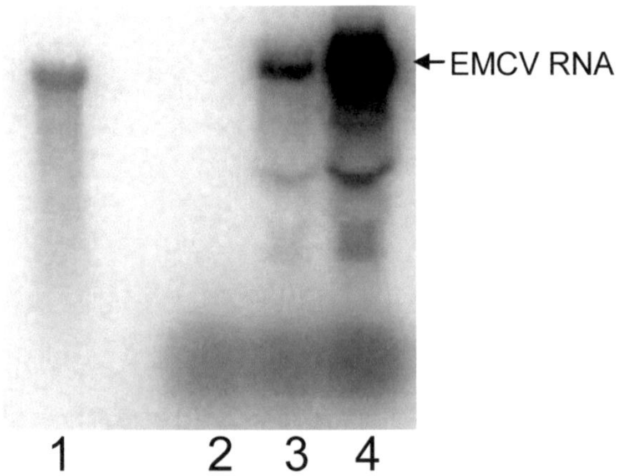

Figure 5. Dialysis enhances replication of EMCV RNA.
EMCV RNA was incubated in the HeLa cell extract with 32P CTP by the batch system (lane 3) or the dialysis system (lane 4).
Lane2: no RNA was Incubated in the batch system. After incubation, RNA was purified, resolved by gel, and detected by autoradiography. Lane 1: in vitro-synthesized EMCV RNA

Furthermore, synthesis of EMCV from DNA templates *in vitro* is now possible (Figure 3). When a plasmid or a PCR product harboring the full-length cDNA of EMCV in the T7 promoter /terminator unit is incubated in the HeLa extract-derived cell-free protein synthesis system supplemented with T7 RNA polymerase, EMCV is progressively produced, thereby circumventing the handling the easily degradable viral RNA (7). This coupled system for the EMCV synthesis provides an opportunity to study the selectiveness of the RNA into the viral particle. Two forms of EMCV RNA are synthesized in the DNA-programmed system: the RNA transcribed from the plasmid by T7 RNA polymerase and the RNA amplified by the viral RNA-dependent RNA polymerase. It is thus imperative to determine which RNA form is incorporated into the EMCV particles. To this end, RNA was purified from EMCV particles from the incubated mixture, and sequencing of the RNA revealed that the EMCV RNA transcribed by the virally encoded RNA-dependent RNA polymerase was predominantly incorporated into the EMCV particle even in the presence of a larger amount of the EMCV RNA transcribed by T7 RNA polymerase from the plasmid (7). This work is a good example that shows the usefulness of the cell-free system for the study of the RNA virus.

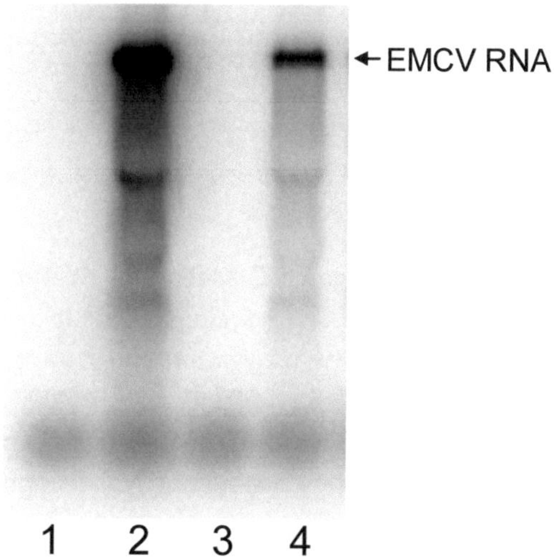

Figure 6. A ribozyme enables replication of synthetic EMCV RNA.
Genomic EMCV RNA (lane 2), synthetic EMCV RNA without (lane 3) or with (lane 4) a ribozyme sequence at the 5′ end was incubated in the HeLa cell extract as in Figure 5 by the dialysis system. Lane 1: no RNA was Incubated. After incubation, RNA was purified, resolved by gel, and detected by autoradiography.

5. Concluding remarks

As discussed in this chapter, the human cells-derived *in vitro*-protein expression systems are unique with fascinating values than compared with other cell-free systems. However, the extract-dependent system contains unknown substances, and proteases in the extract are particularly problematic, since synthesized proteins would be degraded. In this regard, the PURE system (27), which comprises purified translation factors and ribosomes from *E. coli* is an ideal system, and a humanized PURE system should be developed as soon as possible.

Author details

Kodai Machida
Department of Materials Science and Chemistry, Graduate School of Engineering, University of Hyogo, Himeji, Japan
Molecular Nanotechnology Research Center, Graduate School of Engineering, University of Hyogo, Himeji, Japan

Mamiko Masutan
Department of Materials Science and Chemistry, Graduate School of Engineering, University of Hyogo, Himeji, Japan

Hiroaki Imataka*
Department of Materials Science and Chemistry, Graduate School of Engineering, University of Hyogo, Himeji, Japan
Molecular Nanotechnology Research Center, Graduate School of Engineering, University of Hyogo, Himeji, Japan
RIKEN Systems and Structural Biology Center, Tsurumi-ku, Yokohama, Japan

6. References

[1] Kiga D, Sakamoto K, Kodama K, Kigawa T, Matsuda T, Yabuki T, et al. An engineered Escherichia coli tyrosyl-tRNA synthetase for site-specific incorporation of an unnatural amino acid into proteins in eukaryotic translation and its application in a wheat germ cell-free system. Proc Natl Acad Sci U S A 2002;99(15):9715-20.

[2] Kigawa T, Yabuki T, Matsuda N, Matsuda T, Nakajima R, Tanaka A, et al. Preparation of Escherichia coli cell extract for highly productive cell-free protein expression. J Struct Funct Genomics 2004;5(1-2):63-8.

[3] Endo Y, Sawasaki T. High-throughput, genome-scale protein production method based on the wheat germ cell-free expression system. J Struct Funct Genomics 2004;5(1-2):45-57.

[4] Craig D, Howell MT, Gibbs CL, Hunt T, Jackson RJ. Plasmid cDNA-directed protein synthesis in a coupled eukaryotic in vitro transcription-translation system. Nucleic Acids Res 1992;20(19):4987-95.

[5] Mikami S, Kobayashi T, Yokoyama S, Imataka H. A hybridoma-based in vitro translation system that efficiently synthesizes glycoproteins. J Biotechnol 2006;127:65-78.

[6] Mikami S, Kobayashi T, Masutani M, Yokoyama S, Imataka H. A human cell-derived in vitro coupled transcription/translation system optimized for production of recombinant proteins. Protein Expr Purif 2008;62(2):190-8.

[7] Kobayashi T, Nakamura Y, Mikami S, Masutani M, Machida K, Imataka H. Synthesis of encephalomyocarditis virus in a cell-free system: from DNA to RNA virus in one tube. Biotechnol Lett 2012;34(1):67-73.

[8] Helenius A, Aebi M. Roles of N-linked glycans in the endoplasmic reticulum. Annu Rev Biochem 2004;73:1019-49.

[9] Lowe JB, Marth JD. A genetic approach to Mammalian glycan function. Annu Rev Biochem 2003;72:643-91.

* Corresponding Author

[10] Apweiler R, Hermjakob H, Sharon N. On the frequency of protein glycosylation, as deduced from analysis of the SWISS-PROT database. Biochim Biophys Acta 1999;1473(1):4-8.

[11] Walter P, Blobel G. Preparation of microsomal membranes for cotranslational protein translocation. Methods Enzymol 1983;96:84-93.

[12] Mikami S, Masutani M, Sonenberg N, Yokoyama S, Imataka H. An efficient mammalian cell-free translation system supplemented with translation factors. Protein Expr Purif 2006;46:348-57.

[13] Lasky LA, Groopman JE, Fennie CW, Benz PM, Capon DJ, Dowbenko DJ, et al. Neutralization of the AIDS retrovirus by antibodies to a recombinant envelope glycoprotein. Science 1986;233(4760):209-12.

[14] Hinnebusch A. Mechanism and regulation of initiator methionyl-tRNA binding to ribosomes. Cold Spring Harbor, N.W.: Cold Spring Harbor Laboratory; 2000.

[15] Kaufman RJ. Regulation of mRNA translation by protein folding in the endoplasmic reticulum. Trends Biochem Sci 2004;29(3):152-8.

[16] Nonato MC, Widom J, Clardy J. Crystal structure of the N-terminal segment of human eukaryotic translation initiation factor 2alpha. J Biol Chem 2002;277(19):17057-61.

[17] Ramelot TA, Cort JR, Yee AA, Liu F, Goshe MB, Edwards AM, et al. Myxoma virus immunomodulatory protein M156R is a structural mimic of eukaryotic translation initiation factor eIF2alpha. J Mol Biol 2002;322(5):943-54.

[18] Carroll K, Elroy-Stein O, Moss B, Jagus R. Recombinant vaccinia virus K3L gene product prevents activation of double-stranded RNA-dependent, initiation factor 2 alpha-specific protein kinase. J Biol Chem 1993;268(17):12837-42.

[19] Novoa I, Zeng H, Harding HP, Ron D. Feedback inhibition of the unfolded protein response by GADD34-mediated dephosphorylation of eIF2alpha. J Cell Biol 2001;153(5):1011-22.

[20] Mikami S, Kobayashi T, Machida K, Masutani M, Yokoyama S, Imataka H. N-terminally truncated GADD34 proteins are convenient translation enhancers in a human cell-derived in vitro protein synthesis system. Biotechnol Lett 2010;32(7):897-902.

[21] Palmenberg AC. Proteolytic processing of picornaviral polyprotein. Annu Rev Microbiol 1990;44:603-23.

[22] Palmenberg AC. In vitro synthesis and assembly of picornaviral capsid intermediate structures. J Virol 1982;44(3):900-6.

[23] Arnold E, Luo M, Vriend G, Rossmann MG, Palmenberg AC, Parks GD, et al. Implications of the picornavirus capsid structure for polyprotein processing. Proc Natl Acad Sci U S A 1987;84(1):21-5.

[24] Kobayashi T, Mikami S, Yokoyama S, Imataka H. An improved cell-free system for picornavirus synthesis. J Virol Methods 2007;142(1-2):182-8.

[25] Svitkin YV, Sonenberg N. Cell-free synthesis of encephalomyocarditis virus. J Virol 2003;77(11):6551-5.

[26] Kobayashi T, Machida K, Mikami S, Masutani M, Imataka H. Cell-free RNA replication systems based on a human cell extracts-derived in vitro translation system with the encephalomyocarditisvirus RNA. J Biochem 2011;150(4):423-30.
[27] Shimizu Y, Inoue A, Tomari Y, Suzuki T, Yokogawa T, Nishikawa K, et al. Cell-free translation reconstituted with purified components. Nat Biotechnol 2001;19(8):751-5.

Chapter 5

Solid-Phase Cell-Free Protein Synthesis to Improve Protein Foldability

Manish Biyani and Takanori Ichiki

Additional information is available at the end of the chapter

http://dx.doi.org/10.5772/2955

1. Introduction

Proteins are the most abundant molecules in biology which control virtually every biological process on which our lives depend. Therefore, understanding how newly synthesized proteins folds into the correct native structure and achieve their biologically functional states inside the cell is of paramount importance. Most of what is currently known about the process of protein folding has been studied by analyzing proteins outside the cells in a 'dilute solution' under *in vitro* conditions. The pioneering work on the creation of cell-free (*in vitro*) protein synthesis (CFPS) reported by Nirenberg and Matthaei in 1961 has been a powerful and ever expanding tool for large-scale analysis of proteins [1]. In general, these systems are derived from the crude extract of cells engaged in a high rate of protein synthesis and are consist of all the macromolecular components required for translation of exogenous mRNA which are added separately in the system. The cell-free system offer several advantages over traditional cell-based (*in vivo*) systems which are specially not good at making exogenous proteins and those which are toxic to the host cell, undergoes rapid proteolytic degradation or forms inclusion bodies. Cell-free system provides the ability to easily manipulate the reaction components and conditions to favor protein synthesis, decreased sensitivity to product toxicity and suitability for miniaturization and high-throughput applications. With these advantages, there is continuous increasing interest in CFPS system among biotechnologists, molecular biologists and medical or pharmacologists. However, CFPS systems rely on the correct folding of the expressed polypeptide chain into a fully functional three-dimensional protein. Thus 'foldability' of expressed protein in a cell-free system is one of the most challenging conundrums of CFPS science.

The folding issue (misfolding or aggregation) is believed to be caused by excessive collision between growing peptide chain and with other macromolecular components of cell-free system. It is estimated that the total concentration of macromolecules such as proteins, nucleic acids, ribosomes and carbohydrates in the crude cell extract is ranged from 300 to 400 mg/mL that occupy about 30% of total cytoplasmic volume [2]. For easy understanding, if 30% of the volume of a cube is filled with macromolecules of a given size, uniformly distributed, then there is virtually no volume available for additional molecules of a similar size. This leads to 'macromolecular crowding' effect which can result in surprisingly large qualitative and quantitative effects on both the thermodynamic and kinetic of interactions among macromolecules. For example, it can favor the association of macromolecules which may lead to a dramatic acceleration in the rate of protein aggregation (a huge variety of diseases share the pathological feature of aggregated misfolded protein deposits such as formation of amyloid fibrils has a central role in the pathogenesis of Alzheimer disease) [3]. Second, crowding also limits the diffusion of molecules that limits the conformational flexibility of growing polypeptide chains, adding complexity to folding and multimerization reactions. Although CFPS is routinely carried out in relatively dilute solutions but yet the commonly used CFPS systems are estimated with a relatively crowding environments containing ~5% (w/v) of macromolecules [4]. Very recently, the inhibition of cell-free translation of *Rluc* mRNA was confirmed under macromolecular crowding conditions created by adding various biocompatible crowding agents. Interestingly, these crowding agents were observed to show an opposite effects on cell-free transcription reactions [4]. This study confirms that a macromolecule crowding may lead to terminal misfolding and therefore determine the folding rates. Thus protein folding which is crucial to the function of proteins requires controlled handling of translation reaction in CFPS system. In this stream, consideration of the protein behavior in their intracellular milieu is crucial. This chapter presents a novel approach, called solid-phase CFPS, which provides mimetic conditions of an intracellular milieu to facilitate efficient cell-free protein translation of more functionally active proteins.

2. Co-translational protein folding: What we can learn?

Protein synthesis is the universal mechanism for translating the genetic information into functional information in all kingdom of life and all synthesized proteins have in common to fold and express their biological activity. The machine which carries the protein synthesis is the ribosome, a large RNA-protein complex. However, the fundamental understanding of how does the ribosome move along an mRNA and how the linear amino acid sequence of a growing polypeptide chain folds correctly into its unique three-dimensional structure is still not completed. It is widely believed that protein folding generally begins during translation on the ribosome, called 'co-translational folding' [5-7]. This implies that the N-terminal part of a growing polypeptide starts its folding as soon as it has been synthesized, prior to the completion of entire polypeptide chain by the ribosome (see Fig.1). The experimental testing of this elegant idea was already begun in the early 1960s and today there is substantial experimental support for the co-translational folding hypothesis. Very recently, an efficient co-translational folding has been demonstrated by using an engineered multidomain fusion protein [8]. In one another study, the folding yield of fluorescent protein was compared between ribosome-released GFP and chemically denatured GFP. The yield of native fluorescent GFP was dramatically higher with co-translational folding [9]. Although encouraging, but yet many details of co-translational folding pathway remain unanswered. For example, since the fact that the polypeptide synthesis requires many seconds (50-300 residues/min) and the folding occurs in much less than one second (or microsecond-level), there must be formation of compact structures and/or intermediates in the process of protein synthesis. So, what types of structures are these and how they effects on the folding efficiency of newly synthesized protein is still remain elusive.

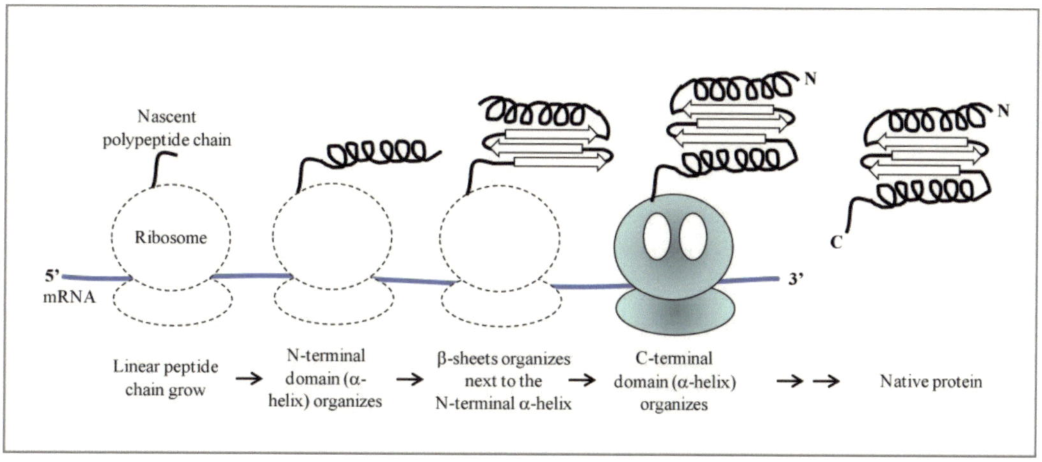

Figure 1. A cartoon representation of 'co-translational folding' of a growing polypeptide chain on the ribosome.

The ribosome serves as a platform for co-translational folding. A crucial process is the decision whether the folding occurs in the cytosol or across the membranes (eukaryotic endoplasmic reticulum (ER) membrane or bacterial plasma membrane). Eukaryotic co-translational protein translocation involves the interaction of signal recognition particle (SRP) with ribosomes. The SRP recognize the hydrophobic signal sequence at the N termini of nascent peptide chains as they emerge from the exit tunnel of ribosome and then SRP-RNC (ribosome–nascent chain) complex interacts with the ER membrane-bound SRP receptor to delivers nascent peptide chain to the ER membrane. This process slowing down chain elongation and lead to a transient arrest of translation. Once ribosome engages a proteinaceous channel located at the ER membrane, only then protein synthesis is resumed and nascent protein are co-translationally injected into the ER lumen. So, what we understand that slowing down the translation rate (as a result of co-translational process) may improve the folding efficiency of newly synthesized proteins. It has been observed that protein synthesis speed is faster in bacteria than in eukaryotes. In *E. coli,* polypeptide synthesis rates vary from 10 to 20 amino acids per second [10] but it is considerably slower (3 to 8 amino acids per second) in the eukaryotes [11]. Presumably, this might be the reason why the eukaryotic cytosol appears to be highly capable of folding proteins efficiently (as a result of co-translational folding) whereas folding of protein is delayed relative to their synthesis in the bacterial cell. It is recently highlighted that a single codon mutation in mRNA that alters the translation rate can lead to a dramatic increase in the folding yield [12]. Thus, the speed of protein synthesis can affect protein folding pathways. And if this is true, then controlling the polypeptide synthesis rate would be promising step to improve the protein foldabiltiy in the CFPS systems. Since both the ribosomes and mRNA templates in the CFPS are not in a stationary mode (as they are in cell-based system represented by endoplasmic reticulum membrane-bound ribosome), providing a similar environment by introducing solid-phase chemistry would help to create co-translational protein folding in the CFPS systems.

3. Solid-phase versus solution-phase chemistry for protein synthesis

Solution dynamics (representing diversity of molecular conformations and motion) of biological macromolecules (e.g., DNAs, mRNAs) has been described by using nanosecond molecular dynamics or X-ray scattering approaches [13,14]. These studies suggest conformational variation including semi-stable or unstable structures having short life times is a general functional feature of these macromolecules and this is profoundly influenced by their environment, such as small changes in the concentration of solutes or salts can radically alter the properties of DNA/mRNA in the solution. These dynamics, such as spatial and temporal dynamic of mRNA movements that undergoes many conformational rearrangements and so an integral part of cell-based protein synthesis, however, may not require in the cell-free systems and thus should be avoided in the cell-free reactions. Secondly, exogenous mRNAs are extremely labile in nature and thus are apt to be degraded by contaminating nucleases that are inherently present in the crude cell extracts and thus the protein synthesis reaction is inhibited over time. Third, since CFPS carried out in relatively dilute concentrations, the ribosome turnover is likely less compared with the cell-based system. In order to exploit these issues, here we introduced solid-phase chemistry for the CFPS systems where the diffusional migration of key molecules (e.g., mRNAs or ribosomes) is restricted in a defined area to improve the positive reactions in a pseudo-first order fashion (see Fig.2).

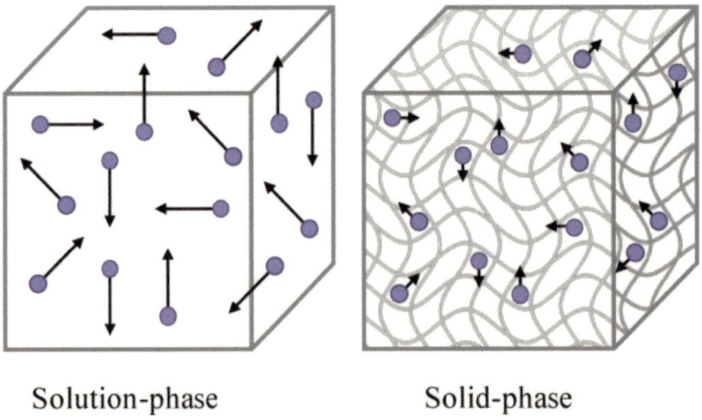

Figure 2. A schematic drawing of diffusional migration in solution-phase and solid-phase reactions. Blue circle represent the CFPS reactants.

Compared with solution-phase reaction, in which the reactants of CFPS are dispersed in a dilute solution, advantages of solid-phase CFPS reaction includes: (i) improved stability: the boundaries stabilized and protect biomolecules by capping the free terminal ends against nucleases degradation; (ii) higher local concentration: the local concentration of the reactants can be greatly increased in solid-phase, a condition that cannot be realized in the solution-phase because of the extra volume of the solvent and the fixed solubility of template DNAs/mRNAs. For example, ribosome-turnover can be increased to find its next substrate in solid-phase reaction; (iii) post-reaction steps: it become easier to perform purifications or remove excess reactant or byproducts from the reaction; (iv) co-translational folding: it mimic the cell-bases system by introducing a diffusion barrier which significantly reduces the reaction rate and improve the co-translational folding. A schematic drawing of solid-phase CFPS is outlined in Fig.3 and compared with solution-phase CFPS and cell-based system. Here, we should recall that protein synthesis is compartmentalized in the cell-based system and secretory/integral proteins being synthesized on endoplasmic reticulum (ER) by trafficking of the ribosome and mRNA from the cytoplasm to the ER membrane. Therefore, solid-phase CFPS where mRNAs are immobilized on a solid surface provides the similar environment with the cell-based system by controlling the reactions in a similar stationary mode using surface-bound mRNA, and this may help to direct protein folding.

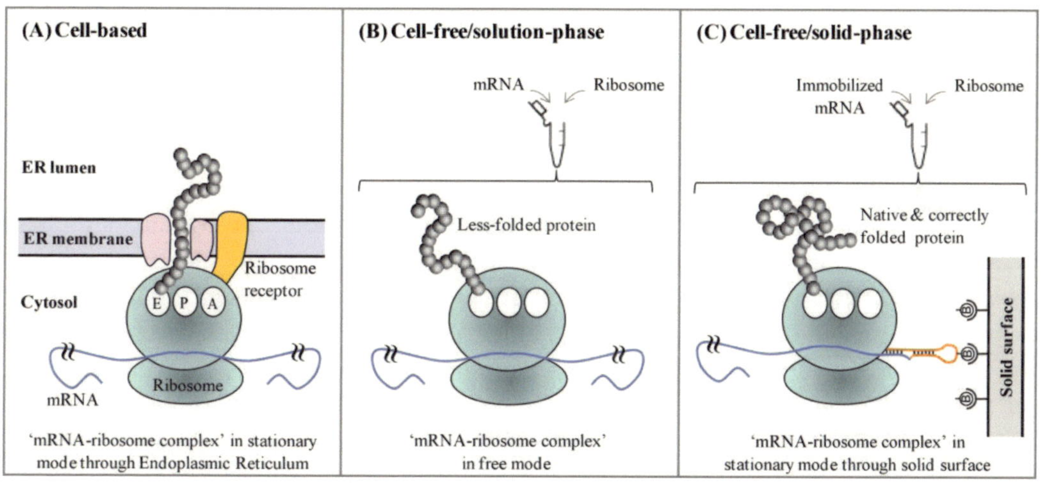

Figure 3. Schematic representation of protein synthesis *in vivo* (A) conventional in vitro solution-phase (B) and novel in vitro solid-phase (C).

The solid-phase approach was first invented by Bruce Merrifield in 1963 in an effort to overcome difficulties inherent to the liquid-phase synthesis of peptide [15]. Later, the immobilization of biomolecule and synthetic solid-phase approaches have been successfully aided research for a widespread applications for both pre-translated biomolecules such as RNA/DNA and post-translated biomolecules such as protein/enzymes including SNP genotyping [16], DNA amplification [17], differential display [18], *in vitro* transcription [19], immunoassay [20,21], and others while promoted the development of microfabrication [22,23], high-throughput screening and automation strategies in many areas including proteomics. Very recently, a hydrogel-based system was introduced that improved the efficiency of CFPS up to 300 times than solution phase-based system [24].

The simplest method for immobilization of biomolecules is physical adsorption between the molecule of interest, e.g., protein, and solid surface [25]. However a more stable and reliable mean of immobilization is a bonding or linkage between the molecule of interest and molecules of the solid support [26]. To date, several methods have been reported to bind the functional biomolecules with ligands onto glass, agarose bead gels and magnetic particles. Among these, the covalent nature bonding affinity has advantageous over non-covalent bonding in the ability to orient the immobilized molecule in a defined and precise fashion for forthcoming reactions. The affinity of biotin for streptavidin is one of the strongest and most stable known in biochemistry [27]. Moreover, a wide range of immobilizing materials and binding modes allows a great deal of flexibility in order to design a specific bond with specific physical and chemical properties such as charge distribution, hydrophobic/hydrophilic, etc. In this chapter, we highlight our new approach of solid-phase protein synthesis to improve the stability and foldability of CFPS systems.

4. General concepts for solid-phase CFPS

In order to exploit the above issues, a novel solid-phase CFPS was described to produce proteins in their native folded-state which is schematically outlined in Fig. 4 [28]. The requires the template (mRNA) in a stationary phase, which is achieved by immobilizing the mRNA molecules to a solid-surface prior to translation. In order to perform solid-phase translation, the immobilization of mRNA must satisfy several requirements: (i) mRNAs should be attached efficiently to the solid surface via a 3′-UTR end linkage, (ii) the integrity of the mRNAs should not be affected by immobilization, (iii) the availability of the free 5′-end of the mRNA must be sufficient for translation and (iv) the properties of the solid-surface must be compatible with translation. These are achieved by coupling the mRNA of interest to a solid surface via ligation to a synthetic biotinylated DNA oligomer which is then immobilized to streptavidin-coated paramagnetic beads. An efficient ligation is an essential part of solid-phase translation and for this purpose we have engineered a synthetic linker-DNA molecule (see Fig. 4A). To perform an efficient ligation between the mRNA and linker-DNA molecules, the 3′-ends of the mRNAs are first hybridized to the linker-DNA and then incubated with T4 RNA ligase. This reaction is efficient even at low concentrations of substrates as it is based on quasi-intramolecular ligation. In the next step of solid-phase translation, the bead-bounded mRNA molecules are incubated in a cell-free translation system (Fig. 4B).

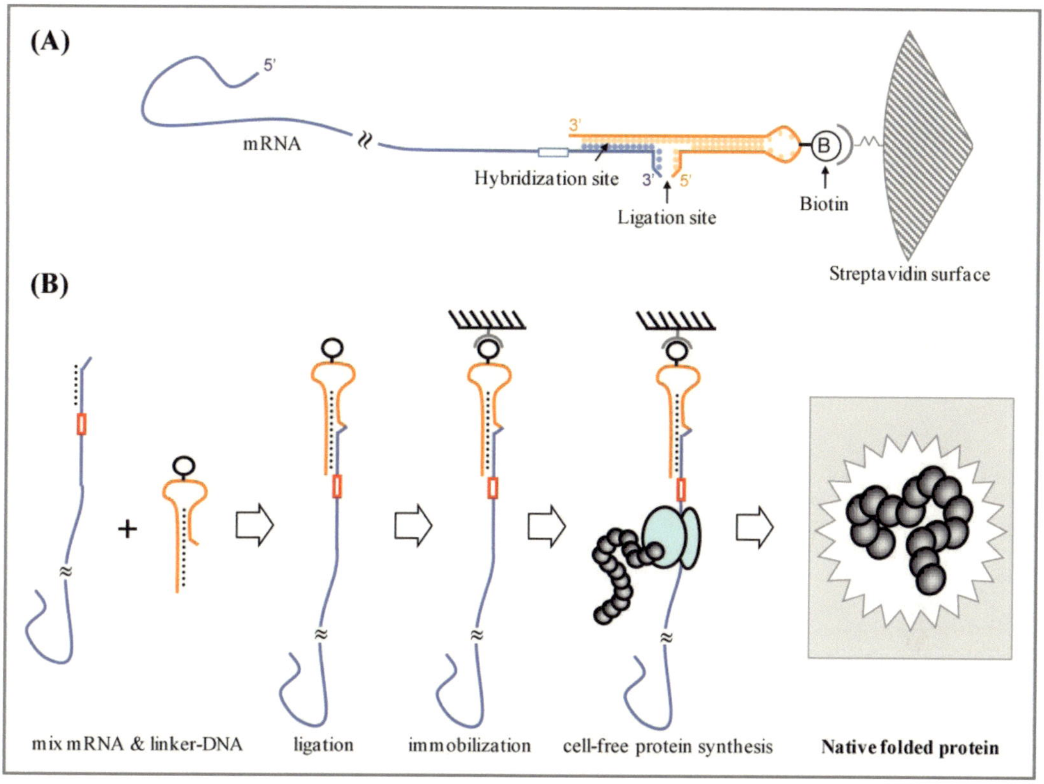

Figure 4. Schematic representation of a novel solid-phase cell-free protein synthesis for synthesizing native and correctly folded protein.

5. Yield of natively folded proteins by solid-phase CFPS

To demonstrate the performance of solid-phase translation system, FP (fluorescent proteins: GFP, green fluorescent protein and mCherry) was chosen as the model proteins. A T7 promoter driven DNA template encoding mCherry with a stop codon was constructed and amplified with biotinylated primer. The PCR products were then immobilized onto streptavidin-coated paramagnetic beads. Following cell-free couple transcription/translation reaction, the beads were separated and the supernatant was analyzed by native SDS-PAGE. To compare the performances of solid-phase and solution-phase systems, an identical quantity of free PCR products without immobilization was processed in parallel. The original fluorescence of the folded mCherry protein was successfully resolved by SDS-PAGE as a major band of ~28 kDa (see Fig. 5A). The RFU (relative fluorescence units) values representing the foldability of mCherry bands were monitored by a fluorescence imager. The average results obtained by three successive experiments clearly show that synthesis of mCherry using novel solid-phase system was at above 2-fold of the solution-phase system (Fig. 5B).

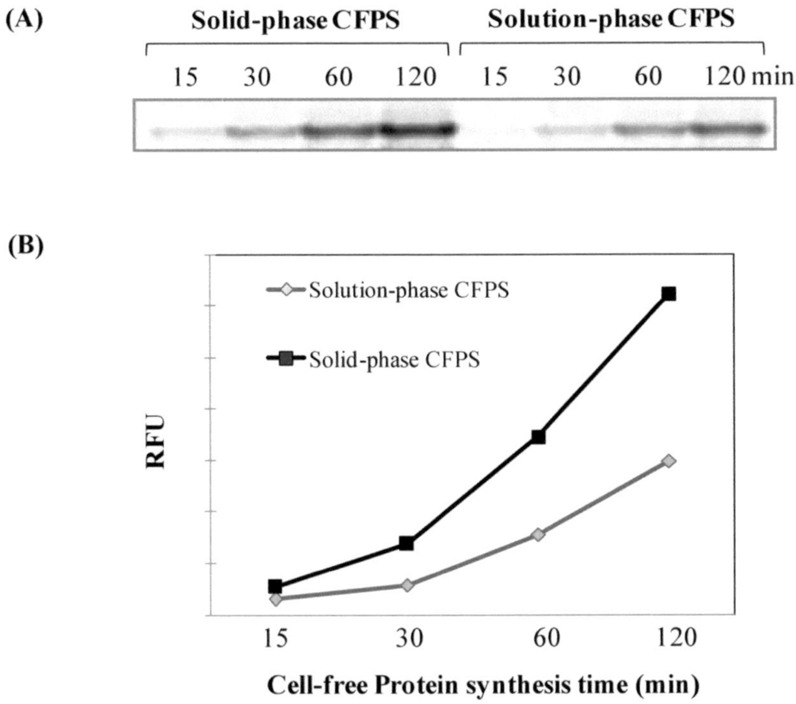

Figure 5. Comparison of the protein synthesis of correctly folded mCherry by solid-phase and solution-phase CFPS. A native SDS-PAGE analysis and quantitative measurements (B).

Indeed, it was surprising to see that by simply converting the free DNA template to surface-bounded template, the efficiency of protein synthesis using coupled transcription/translation system was much improved. To understand this further, we studied the underlying mechanisms by investigating cell-free translation separately. For this purpose, a T7 promoter driven mRNA template encoding GFP with a stop codon and short stretch of complementary sequence of linker-DNA at the 3′-terminus was constructed (as shown partly in Fig.4A). This template was then ligated to linker-DNA and immobilized onto streptavidin-coated paramagnetic beads. Following cell-free translation in a wheat germ-based system, the beads were separated and the supernatant was analyzed quantitatively by SDS-PAGE and qualitatively (i.e., correct folding) by a fluorescence microplate reader. To compare the performances of solid-phase and solution-phase systems, an identical quantity of free mRNA-template without ligation or immobilization was processed in parallel. To quantitatively compare the production between the solid- and solution-phase methods, GFP was expressed using fluorescently labeled lysine residues. Translated products were heated at 70°C for 5 min for complete denaturation and removal of the original fluorescence of the folded GFP protein, and resolved by SDS-PAGE. Heat-denatured (non-fluorescent) GFP migrates as a major band of about 27 kDa (Fig.6A, right two lanes). The intensity of FluoroTect labeled GFP bands were monitored by a fluorescence imager. The average results obtained by four successive experiments clearly show that production of GFP using our solid-phase system was at about 15% of the levels of the liquid-phase system (Fig.6B inset, white columns). However, the quality analysis, i.e., foldability, of the GFP, for these two systems obtained by measuring the intensity of original green fluorescence, (Fig.6B inset, gray columns) showed similar results. The RFU (relative fluorescence units) values representing the foldability of GFP were directly measured using a fluorescence microplate reader, and for the solid-phase system was about 80% of the liquid-phase system. This suggests that although the production of GFP using the solid-phase approach is considerably less compared with the liquid-phase method, the proteins produced in the solid phase are up to four-fold more biologically active after normalization (Fig.6B). To confirm this finding, the solid-phase products were removed from the beads and then analyzed together with solution-phase products by SDS-PAGE. The results showed a 37 kDa GFP product from the solid phase reaction, which is shifted upwards from the denatured position predicted for its theoretical mass (27 kDa) due to its native folding (Fig.6A, left two lanes).

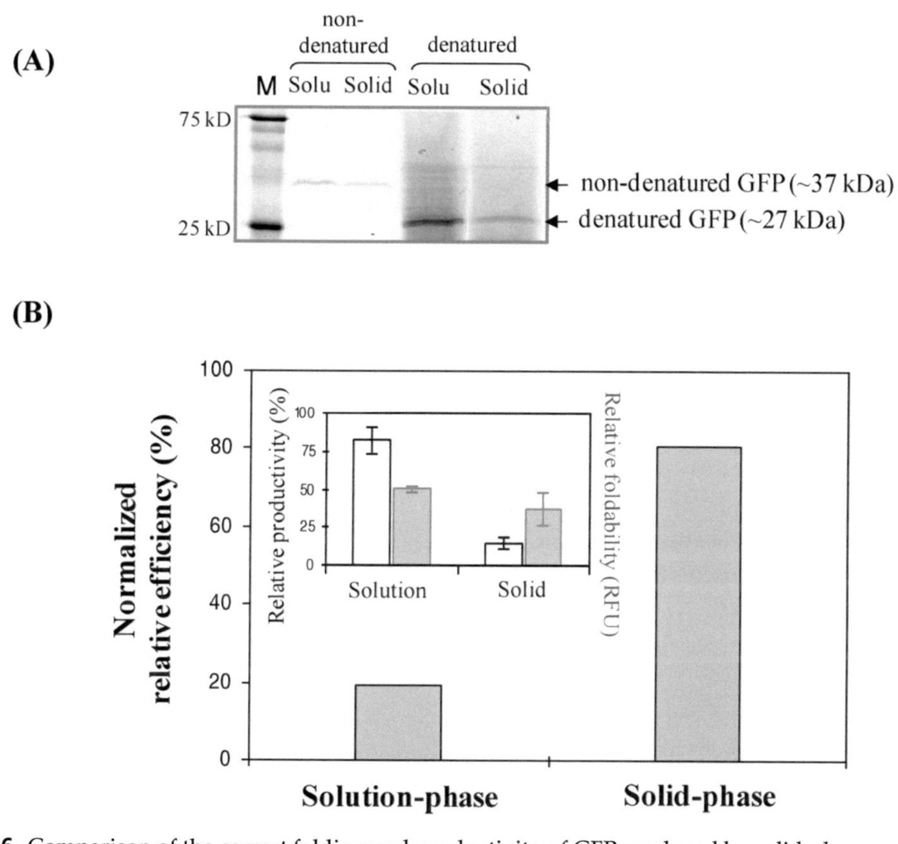

Figure 6. Comparison of the correct folding and productivity of GFP produced by solid-phase and solution-phase CFPS systems. (A) SDS-PAGE of non-denatured (folded) GFP (leftmost two lanes) and denatured GFP (rightmost two lanes). (b) Quantitative measurements of the relative efficiency of Solid versus Solution-phase in terms of ratio values were plotted after recombining the productivity (white column) and foldability (grey column) performance (shown in inset) from the solution or solid-phase systems. All plots and error bars represent average and standard coefficient values of more than four independent experiments. M, molecular weight markers.

6. Conclusion and future perspective

This chapter described a novel solid-phase cell-free translation system in which template molecules (DNAs/mRNAs) were captured onto solid-surfaces to simultaneously induce co-translational folding and synthesize proteins in a more native-state form. A newly constructed biotinylated linker-DNA is ligated to the 3′ ends of the mRNA molecules to attach the mRNA-template on a streptavidin-coated surface and further to enable the subsequent reactions of cell-free translation on surface. The protein products are therefore directly synthesized onto solid-surfaces and furthermore discovered to adopt a more native state with proper protein folding and enough biological activity compared with conventional solution-phase approaches. The approach described in this chapter may enables to embrace the concept of the transformation of 'DNA-to-Protein microarrays' using solid-phase cell-free protein synthesis system and thus to the development of high-throughput, CFPS platform to the field of functional proteomics.

Author details

Manish Biyani
Department of Bioengineering, The University of Tokyo, Bunkyo-ku, Tokyo, Japan
Japan Science and Technology Agency, CREST, Chiyoda, Tokyo, Japan
Department of Biotechnology, Biyani Group of Colleges, R-4, Sector No 3, Vidhyadhar Nagar, Jaipur, India

Takanori Ichiki
Department of Bioengineering, The University of Tokyo, Bunkyo-ku, Tokyo, Japan
Japan Science and Technology Agency, CREST, Chiyoda, Tokyo, Japan

7. References

[1] Nirenberg MW, Matthaei JH. The dependence of cell-free protein synthesis in E. coli upon naturally occurring or synthetic polyribonucleotides. *Proc Natl Acad Sci U S A.* 1961, 47, 1588-1602.
[2] Ellis RJ. Macromolecular crowding: obvious but underappreciated. *Trends Biochem Sci.* 2001, 26, 597-604.
[3] Munishkina LA, Cooper EM, Uversky VN, Fink AL. The effect of macromolecular crowding on protein aggregation and amyloid fibril formation. *J Mol Recognit.* 2004, 17, 456-464.
[4] Ge X, Luo D, Xu J. Cell-free protein expression under macromolecular crowding conditions. *PLoS One* 2011, 6, e28707.
[5] Fedorov AN, Baldwin TO. Cotranslational protein folding. *J Biol Chem.* 1997, 272, 32715–32718.
[6] Komar AA. A pause for thought along the co-translational folding pathway. *Trends Biochem Sci* 2009, 34, 16–24.

[7] Kramer G, Boehringer D, Ban N, Bukau B. The ribosome as a platform for co-translational processing, folding and targeting of newly synthesized proteins. *Nat Struct Mol Biol.* 2009; 16:589–597.

[8] Han Y, David A, Liu B, Magadán JG, Bennink JR, Yewdell JW, Qian SB. Monitoring cotranslational protein folding in mammalian cells at codon resolution. *Proc Natl Acad Sci U S A.* 2012, 109, 12467-12472.

[9] Ugrinov KG, Clark PL. Cotranslational folding increases GFP folding yield. *Biophys J.* 2010, 98, 1312-1320.

[10] Liang, ST, Xu, YC, Dennis, P, Bremer, H. mRNA composition and control of bacterial gene expression. *J. Bacteriol.* 2000, 182, 3037–3044.

[11] Mathews, MB, Sonenberg, N, Hershey, JWB. Origins and principles of translational control. In *Translational Control of Gene Expression* (Sonenberg, N, Hershey, JWB, Mathews, MB, eds), 2000, 1–31, Cold Spring Harbor Laboratory Press, Cold Spring Harbor, NY.

[12] O'Brien EP, Vendruscolo M, Dobson CM. Prediction of variable translation rate effects on cotranslational protein folding. *Nat Commun.* 2012, 3, 868.

[13] Biyani M, Nishigaki K. Single-strand conformation polymorphism (SSCP) of oligodeoxyribonucleotides: an insight into solution structural dynamics of DNAs provided by gel electrophoresis and molecular dynamics simulations. *J Biochem.* 2005, 138, 363-73.

[14] Rambo RP, Tainer JA. Bridging the solution divide: comprehensive structural analyses of dynamic RNA, DNA, and protein assemblies by small-angle X-ray scattering. *Curr Opin Struct Biol.* 2010, 20, 128-137.

[15] Merrifield, RB. Solid Phase Peptide Synthesis. I. The Synthesis of a Tetrapeptide. *Journal of the American Chemical Society.* 1963, 85, 2149-2154.

[16] Lockley AK, Jones CG, Bruce JS, Franklin SJ, Bardsley RG. Colorimetric detection of immobilised PCR products generated on a solid support. *Nucleic Acids Res.* 1997, 25, 1313-1314.

[17] Adessi C, Matton G, Ayala G, Turcatti G, Mermod JJ, Mayer P, Kawashima E. Solid phase DNA amplification: characterisation of primer attachment and amplification mechanisms. *Nucleic Acids Res.* 2000, 28, e87.

[18] Ståhl S, Odeberg J, Larsson M, Røsok O, Ree AH, Lundeberg J. Solid-phase differential display and bacterial expression systems in selection and functional analysis of cDNAs. *Methods Enzymol.* 1999, 303, 495-511.

[19] Marble HA, Davis RH. RNA transcription from immobilized DNA templates. *Biotechnol Prog.* 1995, 11, 393-396.

[20] Catt K, Niall HD, Tregear GW. Solid phase radioimmunoassay. *Nature.* 1967, 213, 825-827.

[21] Butler JE. Solid supports in enzyme-linked immunosorbent assay and other solid-phase immunoassays. *Methods.* 2000, 22, 4-23.

[22] Schena M, Shalon D, Davis RW, Brown PO. Quantitative monitoring of gene expression patterns with a complementary DNA microarray. *Science.* 1995, 270, 467-470.

[23] Schena M, Shalon D, Heller R, Chai A, Brown PO, Davis RW. Parallel human genome analysis: microarray-based expression monitoring of 1000 genes. *Proc Natl Acad Sci U S A*. 1996, 93, 10614-10619.

[24] Park N, Um SH, Funabashi H, Xu J, Luo D. A cell-free protein-producing gel. *Nat Mater.* 2009, 8, 432-437.

[25] Andrade JD, Hlady V. Protein adsorption and materials biocampatibility: a tutorial review and suggested hypotheses. In: Dusek K, editor. *Advances in polymer science.* Berlin: Springer, Heidelberg, 1986.

[26] Weetall HH. Preparation of immobilized proteins covalently coupled through silane coupling agents to inorganic supports. *Appl Biochem Biotechnol.* 1993, 41, 157-188.

[27] Laitinen OH, Nordlund HR, Hytönen VP, Kulomaa MS. Brave new (strept)avidins in biotechnology. *Trends Biotechnol.* 2007, 25, 269-77.

[28] Biyani M, Husimi Y, Nemoto N. Solid-phase translation and RNA-protein fusion: a novel approach for folding quality control and direct immobilization of proteins using anchored mRNA. *Nucleic Acids Res.* 2006, 34, e140.

Section 4

Translational Control and Protein Synthesis

Cumulative Specificity: A Universal Mechanism for the Initiation of Protein Synthesis

Tokumasa Nakamoto and Ferenc J. Kezdy

Additional information is available at the end of the chapter

http://dx.doi.org/10.5772/51179

1. Introduction

The main problem in the initiation of protein synthesis is the determination of how the ribosome recognizes and binds to the initiation site (IS) of the mRNA. There are currently three major hypotheses that address this problem, all differently. The Shine-Dalgarno (SD) hypothesis for prokaryotes proposed in 1974, postulates that the IS is selected by base pairing of a segment of the 3′ end of the 16S rRNA of the small ribosomal subunit and a complementary segment in the leader sequence of the IS [1]. The scanning hypothesis for eukaryotes proposed in 1978, postulates that the 40S ribosome initiation complex recognizes and binds to the 5′ end of the mRNA and scans the IS until it finds the initiator codon [2]. The cumulative specificity (CS) hypothesis has its origin in a 1966 proposal that provided an essentially unique accessibility of the IS proposal for prokaryotes [3], but with a number of recent modifications, it evolved into its current form in 2007 [4]. The CS mechanism postulates that the IS of the mRNA is selected by incremental ribosomal binding of the IS, the ribosomal binding subsites interacting with their respective IS subsites, one or a few subsites at a time.

An important aspect of any good hypothesis is its ability to stimulate research. The long tenure of the SD and the scanning hypotheses as the bases for numerous researches attest to the credibility and appeal of the two hypotheses. However, with manifold increase of knowledge in the field, evaluation of the current hypotheses is timely, especially the SD and the scanning hypotheses, which were proposed so long ago. Repeating, the major hypotheses are the SD proposal for prokaryotes, the scanning mechanism for eukaryotes, and the CS mechanism, for both prokaryotes and eukaryotes. On another postulated mechanism, the internal ribosome entry site (IRES) for eukaryotes [5], there is some question whether it and the scanning mechanism are indeed distinct and unconnected [6] so the former will not be considered here as a major mechanism.

The model IS of the E. coli mRNA, generated by computer analysis from 68 non-identical IS sequences, consists of 46-48 nucleotides with preferred bases (recognition elements) in given positions, but without a specific base in any given position except for the initiator codon [7]. This means that the IS surface is extensive, nonrigid, complex and the IS is a non-unique sequence. The IS of the eukaryotic mRNA has characteristics similar to those obtained for *E. coli*, although not documented as convincingly as in *E. coli* [8]. The extensiveness, nonrigidity, and complexity of the IS of the mRNA would make its binding to the ribosome — that is, the perfect meshing together of the large and complex surface of the IS and that of the ribosome — unlikely to occur in a single collision. Rather, the binding of the IS of the mRNA by the ribosome should occur by one or severable sub-segments at a time as proposed by the CS model reaction, which was originally developed to account for the specificity of substrate binding to the HIV Type 1 and 2 proteases [9]. The CS reaction was then proposed as a paradigm for the initiation of protein synthesis [4]. The main difference between the two above binding reactions is that in one case, an enzyme binds its substrate, and in the other, a ribosome binds a template mRNA.

The specificity sites of the protease substrates consist of a sequence of 6-8 amino acids, and the recognition signals of these substrate sites also consist of preferred recognition elements — viz. amino acids — in given positions [9]. Thus these active sites are also extended, nonrigid and complex and have non-unique sequences of amino acids and are similar in general characteristics to those described above for the IS of mRNA. The CS model reaction assumes that the protease binds the specificity site of the substrate, by a sub-segment or several sub-segments at a time. In the initial collision of the enzyme and substrate, the enzyme first binds to one or more subsites of the substrate. This process is then followed in the protease and protein synthesis reactions by sequential zipper-like juxtapositions and bindings of the appropriate, remaining subsites of the enzyme, or the ribosome, and the substrate (or the mRNA), until completion of the enzyme-substrate (ribosome-mRNA) binding.

A very important feature of the CS mechanism is that it is able to recognize as a signal, not a rigid, immovable molecular structure, but a structure variable within certain limits, namely a sequence of preferred molecular elements in given positions. The consequence of such a heterogeneous, multi-pronged binding is that the ensuing chemical transformation, or the binding of two large surfaces, does not occur with a single reaction rate, which would reflect an all-or-none specificity. Rather, the rate — and hence the specificity — spans over a wide range, favorable interactions at each subsite contributing incrementally to the overall recognition of the substrate active site or of the IS of the mRNA.

Further insights on the initiation mechanism of protein synthesis can be obtained by considering the implications on the general mechanisms of protein synthesis of the evolutionary process and logic [6,10,11]. An important implication of this process and logic is that the general mechanisms of protein synthesis, including the initiation mechanism, are expected to be universal in all domains of life. The universality of the mechanisms of protein synthesis follows from the conservation of the complex protein synthesizing apparatus, exhibiting very similar basic components, consisting of ribosome, mRNA, tRNAs (including an initiator methionyl-tRNA), aminoacyl-tRNA synthetases, a universal genetic code, and numerous other proteins and factors. This review will examine whether the various current hypotheses consider these implications of evolution, along with the characteristics of the IS and other logical points. Such an examination will help to evaluate the comprehensiveness and reasonableness of the initiation mechanisms advanced by the various hypotheses.

2. Body

2.1. Basis for evaluating the hypotheses for the initiation of protein synthesis

The evolutionary logic, previously presented for evaluating the various hypotheses proposed for the mechanism of initiation of protein synthesis [6,10,11], should perhaps rather be thought of as an ensemble of logical considerations gleaned from general knowledge, compiled here to contribute to the evaluation of current hypotheses. We hope that these intuitive and axiomatic considerations do provide a firm foundation for the assertions and conclusions made in this chapter.

As mentioned above, evolutionary logic leads one to conclude that with the conservation of the numerous constituent elements of the complex protein synthetic apparatus, the underlying mechanisms of the synthetic process should also be conserved in all domains of life and are universal. As reviewed later, this conservation also extends in the domains of prokaryotes and eukaryotes to the similarity of mRNA initiation signals and to the ability of the ribosomes to recognize these signals [10,11]. Ribosomes can recognize the mRNA from a different domain of life only if the initiation signals are identical or very closely related and if ribosomes have also conserved their ability to recognize these signals. Taken together, these facts can indicate a conserved initiation mechanism.

The mechanism of initiation of protein synthesis must have evolved gradually and without abrupt changes, from prokaryotes to eukaryotes. Therefore, any hypothesis postulating that this evolution is accompanied by profound changes in the mechanistic characteristics of the ribosomes is rather improbable. This will have bearing in the evaluation of the scanning hypothesis for eukaryotes.

When a hypothesis postulates an exclusive and primary biological pathway — such as the pathway for the initiation of protein synthesis —then the pathway's unique initiation signals and the other essential components of the reaction, must all be obligatory. Moreover, the proposed pathway must be in accord with all experimental observations, such as the characteristics of the initiation sites. These considerations will be central in the evaluation of the SD hypothesis and, to a lesser degree, the other hypotheses.

Evolution is a very efficient process and is not likely to tolerate the conservation of unused information. For instance, the characteristics of the model E. coli initiation site suggests the presence of 46-48 nucleotides with signal character. Thus, any comprehensive initiation mechanism should have an initiation signal that encompasses the entire nucleotide sequence shown to have signal character, which includes essentially the entire IS, even the amino acid coding region. The existence and translation of non-SD and leaderless mRNAs indicate that any hypothesis that requires the recognition of only the leader sequence cannot be a complete hypothesis. A comprehensive initiation mechanism must, therefore, be able to account for the initiation of translation of canonical, as well as, of leaderless mRNAs.

Lastly, as mentioned, the hypotheses will be evaluated on the basis of whether the characteristics of the IS and the dictum of a universal initiation mechanism were considered in formulating each respective initiation mechanism.

2.2. Characteristics of the Initiation Site

The characteristics of the initiation site (IS) were determined by computer analysis of 68 non-identical E. coli ribosome binding site sequences [7]. The analysis generated a model IS containing 46-48 nucleotides that assigns only preferred bases rather than an absolute requirement for a specific base in any given position. The only exception to the preceding is the absolute requirement of the initiator codon in a given position in every IS.

The model IS reveals important characteristics of the IS. Part of the model sequence is complementary to the 3′ of the 16S rRNA, that is, its base frequency profile reflects the SD sequence. Nonetheless, there is no unique initiation sequence or even a unique SD sequence in the IS, only ensembles of preferred bases. The prokaryotic ISs thus constitute a large multiplicity of loosely related nucleotide sequences, and the protein synthesizing system, which recognizes all of them, has broad substrate specificity [4]. Nearly half of the model site covers the amino acid coding region. In other words, it is the entire IS — about one-half of it consisting of the leader region and the other half of the amino acid coding region — that serves as the ribosome recognition or initiation signal. The authors suggested that the finding that the amino acid coding region has recognition features, or signal character, might explain how leaderless mRNAs are recognized by the ribosomes.

The model E. coli IS is a composite of three sequences derived by analysis of the 68 E. coli ribosome binding site sequences using the initiator codon, the SD leader sequence and the CU at the very 3′ end of the IS as the common reference in aligning the IS sequences for analysis. When the initiator codon was the common reference, the model sequence derived showed considerable ambiguity in the area where the SD sequence was expected to be located. When the SD sequence was used as the common reference and the leader region was analyzed, the AUG start codon had a variable locus of 6-9 positions in the 3′ direction from the SD sequence. Examination of the sequences showed that CU seemed to be the 3' end of the model sequence. As with the SD sequence and its variable distance from the AUG codon, analysis found a completely random behavior with the CU as the common reference. A realignment of the sequences and subsequent analysis, however, showed a contiguous and almost identical model sequence between the AUG codon and the assumed CUCG end signal. The best model sequences fragments obtained from the analyses using the three common references described above were then combined to form the composite model E. coli initiation site.

2.3. The conservation of prokaryotic initiation signals and prokaryotic initiation mechanism in eukaryotes

Before discussing specific conservations, it helps to clarify the conclusion that all underlying mechanisms of the protein synthetic process should be conserved in all domains of life because of the conservation of the numerous constituent elements of the complex protein synthetic apparatus. Does this mean that the entire underlying protein synthetic mechanisms are universally conserved? We shall review below that, in addition to conservation of the physical apparatus and the underlying general mechanisms, other aspects of the process, like initiation mechanism features, such as prokaryotic initiation signals with the underlying prokaryotic initiation mechanism, are conserved in eukaryotes. This could also imply that the mechanisms of peptide bond formation, peptide chain elongation and protein chain termination could in all domains of life be identical or very similar to each other.

It may be at present convenient then, when concluding that the initiation mechanism is conserved in eukaryotes, to assume tacitly that the rest of the mechanisms of the protein synthetic pathway are conserved as well. After all, initiation signals alone do not comprise all of the initiation mechanism, and some additional ribosome functions or mechanism are needed to complete the process. In any case, this tacit assumption does not invalidate the experimental observations from which the conclusion of conservation was deduced, nor does it invalidate the deductive process.

The conservation of prokaryotic initiation signals and the prokaryotic initiation mechanism in eukaryotes deduced from experiments using heterologous systems, in which eukaryotic mRNAs were translated in E. coli cell-free systems and prokaryotic mRNAs were translated in eukaryotic cell-free systems, has been reviewed [6,10,11]. The experiments shed light especially on the mechanism of initiation of eukaryotic protein synthesis and provided support for a common or a universal initiation mechanism shared by prokaryotes and eukaryotes. The studies are reviewed, again, with emphasis on the significance of the conservation of initiation signals and the initiation mechanism.

Polypeptides were synthesized in an E. coli cell-free system with poliovirus as messenger whose tryptic digests were found to correspond to tryptic digests of authentic poliovirus proteins [12]. The tobacco mosaic virus (TMV) RNA in an E. coli cell-free system directed the synthesis of several discrete polypeptides, including one similar to TMV coat protein by criteria of polyacrylamide electrophoresis and peptide mapping [13]. Avian myeloblastosis viral RNA was translated by E. coli ribosomes to yield a protein that was antigenically identical to the group-specific antigen 4 of the virus [14]. The preceding experiments clearly show that E. coli, or prokaryotic, ribosomes can recognize eukaryotic viral initiation signals and translate the eukaryotic viral mRNAs. This indicates that the eukaryotic viral mRNAs contain evolutionarily conserved prokaryotic initiation signals since it is unlikely that the two sets of mRNAs would have dissimilar initiation signals and that the same ribosomes would recognize both sets of signals.

Studies were also performed using prokaryotic mRNAs in a eukaryotic cell-free system. Bacteriophage Qβ RNA, a polycistronic prokaryotic messenger, was translated in extracts of Krebs II mouse ascites cell-free system [15]. Viral coat protein was identified as the primary product by co-migration on polyacrylamide gel with authentic coat protein, and by mapping of tryptic digests. A specific mRNA for a structural lipoprotein of E. coli was translated in a wheat germ cell-free system [16]. Eukaryotic ribosomes can thus faithfully translated prokaryotic initiation signals and initiate translation. The translation of prokaryotic mRNAs by eukaryotic ribosomes also means that the initiation signals in prokaryotes and eukaryotes are identical or very similar, and that the eukaryotic ribosomes translate prokaryotic mRNAs by initiating translation with a prokaryotic-like mechanism, or with an evolutionarily conserved prokaryotic mechanism.

Experiments with heterologous systems that were extremely important in establishing the universality of the initiation signals involved the translation of capped prokaryotic mRNAs by eukaryotic ribosomes. In the experiments, λ phage 8S cro mRNA and other λ transcripts that still retained their prokaryotic ISs, when capped in vitro, were found to be translated in a wheat germ cell-free system as efficiently as — or even more efficiently than—naturally capped eukaryotic mRNAs [17,18]. The efficient translation of capped prokaryotic mRNAs by a eukaryotic cell-free system means that prokaryotic initiation signals are equivalent to, or as effective as, the initiation signals of naturally capped eukaryotic mRNAs. Additionally, in light of the cross heterologous translations reviewed above, the preceding can be construed as evidence that prokaryotic initiation signals with the underlying prokaryotic initiation mechanism are conserved in eukaryotes. Thus, the heterologous experiments indicate the conservation of a universal initiation mechanism, or at least, the conservation of a common initiation mechanism in the domains of the prokaryotes and the eukaryotes.

2.4. Ribosomal initiation complex

The ribosome is the predominant constituent of the complex protein synthesizing apparatus. In all domains of life it is composed of two subunits, one small and one large. It was assumed for some time that the subunits of the ribosome functioned in protein synthesis combined as a 70 S ribosome unit. In 1967, however, it was demonstrated that the prokaryotic 70S ribosome operates in a cycle when participating in protein synthesis. The 70S ribosome dissociates into subunits when initiating synthesis, re-associates into the 70S ribosome at the completion of initiation and during polypeptide chain elongation, and again dissociates into subunits when synthesis is complete [19]. Phage f2 RNA as well as poly AUG (1:1:1) in random sequence was shown to stimulate binding of fMet-tRNA to the 30S ribosomal subunits, but not to the 70S ribosome. The presence of the 50S subunit inhibited the binding of fMet-tRNA to the 30S subunit with phage f2 RNA and the poly AGU. It was proposed that the first step in protein synthesis is the formation of a complex consisting of the 30S subunit, mRNA, and fMet-tRNA. Later studies established that the binding reaction required the participation of GTP and 3 specific initiation protein factors [20].

Although the 70S ribosome does not bind fMet-tRNA with f2RNA or polyAGU, it does bind fMet-tRNA with the triplet AUG. This indicates that the mRNA binding site of the 70S ribosome can accommodate a small single triplet and not the larger f2 RNA or polyAGU containing an AUG codon. These observations support the view that the 30S subunit, being only about a third the size of the 70S ribosome, can access the IS more effectively. Another obvious possibility for the lower effectiveness of the 70S ribosome to bind mRNA may be due in part to the shielding by the 50S subunit in the 70S ribosome of the mRNA biding site. A major reason for the difficulty of ribosome binding the IS may be that RNA interacts intra-molecularly so extensively that a ribosome binding to a nucleotide sequence of about 47 nucleotides long, which is the length of the model E. coli mRNA initiation site, is no easy feat. A synthetic RNA polymer containing 4 bases in equal proportions in random sequence has been found to have about 50% of its bases paired [21]. The effectiveness of the small ribosomal subunit to access the IS is probably the primary advantage of the ribosome cycle. Besides, the 70S ribosome is effective and essential for polypeptide chain elongation.

Initiation of eukaryotic protein synthesis follows the same pattern as in prokaryotes with the small ribosome subunit, the 40S ribosome, forming an initiation complex with eukaryotic initiator Met-tRNA that, however, is not formylated. The initiation of the eukaryotic protein synthesizing process requires at least 5 initiation factors and is more complex [22].

2.5. A Cumulative Specificity (CS) reaction model for proteases from human immunodeficiency virus (HIV) Types 1 and 2

The substrate specificities of the aspartyl proteases from HIV Types 1 and 2 were not readily discernible. The scissile bond of the substrate chain is surrounded by at least 3-4 amino acids on each side without any specific sequence or even a specific amino acid in any given position,. To gather some evidence for the basis of the specificity of these enzymes, the frequencies of amino acid distribution at each position surrounding the cleavage site of the of HIV 1 protease substrate were statistically analyzed in 40 substrates with known amino acid sequences [9]. The analysis revealed that certain amino acid residues had quite higher than normal frequencies at three subsites in addition to the positions of the amino acids directly involved in the cleavage, but there was no absolute requirement for any given specific amino acid in any of these positions. Thus, each subsite appeared to have a marked specificity toward some residues, and also a marked negative specificity, since some amino acid residues did not occur at all at those subsites. Inferring that the characteristics of the frequencies of the particular amino acids in given positions are the result of actual molecular interactions between the substrate and the active site of the enzyme, a mechanistic model was proposed to account for the broad specificity of the HIV proteases.

The model postulates that the positioning of the cleavage site with respect to the catalytic groups of the enzyme is the result of the cooperative interaction between the amino acid residues of the substrate subsites and corresponding subsites surrounding the active site of the enzyme. Each of these mutually independent interactions contributes incrementally to the optimization of the positioning of the cleavage site with respect to the catalytic groups and for this reason the model is called the cumulative specificity mechanism. According to this model, the reaction begins with the collision of the enzyme and the substrate, resulting in binding at one of the subsites, presumably the most accessible one and the one with the most favorable interactions. Once the substrate peptide chain is immobilized, this initial binding is then rapidly followed by the independent and sequential interactions with the other substrate subsites. The sequential sub-segment interaction mechanism is most likely, since the peptide chain of the substrate is not rigid and the binding of the enzyme and the subsites could not occur in a single collision.

In summary, the mechanistic model postulates that the catalytically productive positioning of the scissile bond results from the cumulative effect of independent interactions between each substrate side-chain and its respective enzyme subsite. According to this view, none of these individual interactions is absolutely essential as long as the peptide is properly anchored at a sufficient number of adjacent subsites. These subsites should be independent, since; there is no discernible cross-correlation between any pair of amino acids occupying any two subsites. Finally, the concept of negative specificity accounts for the fact that the presence of unfavorable interactions at certain subsites can actually prevent any peptide chain from being a substrate for the enzyme.

The "cumulative specificity model" provides a rational interpretation of the puzzling multiplicity of natural substrates for HIV proteases. The essential features of the model are that no subsite has absolute specificity and that a combination of several mediocre interactions is at least equivalent to the combination of a few strong interactions, as far as the catalytic, as opposed to the binding specificity is concerned. The broad specificity of the HIV- 1 protease appears to follow from its ability to bind productively substrates in which interactions with only a few of the amino acid residues in the subsites need be optimized, that is, the amino acids need to have sufficiently high frequencies.

The analysis, extended to 22 peptide segments cleaved by the HIV 2 protease, delineated marked differences in specificity from that of the HIV 1 enzyme. Since the HIV 1 and 2 proteases are very similar enzymes, both having extended substrate active sites and recognition signals of the sites appear to be preferred amino acids in given positions, it was concluded that the cumulative specificity model was also the mechanism for the HIV 2 protease, as well.

3. Current hypotheses for initiation mechanisms

3.1. Shine-dalgarno hypothesis for prokaryotes

Until recently no controversy existed about the initiation mechanism in prokaryotic protein synthesis. Most of the attention has been focused on the Shine-Dalgarno (SD) hypothesis [1]. It postulates that the initiation site or signal is selected by the base pairing of a nucleotide sequence preceding the initiation codon, currently known as the Shine-Dalgarno (SD) sequence, and a complementary nucleotide sequence at the 3′ end of the 30S ribosome's 16S rRNA (the anti-SD sequence). The proposal that a unique nucleotide sequence base pairing was the basis for IS selection was so attractive that the SD mechanism quickly became accepted even without rigorous proof, and even in the presence of conflicting evidences [23].

The initiation signal of the SD hypothesis, the SD sequence, is composed of about 8 nucleotides [7]. Thus, a sequence of about 8 nucleotides, which is less than 20% of the approximately 47 nucleotides of the E. coli model IS with signal character, raises the evolutionary problem of conservation of about 80% of unused information of initiation signals. Further studies showed that the nucleotide spacing between the SD sequence and the initiator codon can vary as much as 6 to 9 nucleotides [7,24]. All the above raise doubts on whether the SD interaction alone can effectively direct the in-reading-frame binding of the ribosomes to the mRNA.

As discussed before, the essentials of a proposed mechanism for a central biological reaction should be obligatory. In the case of the SD mechanism, the SD sequence, the anti-SD sequence of the 16S rRNA, and the base pairing of the SD sequence with the anti-SD sequence of the ribosome (the SD interaction) should all be obligatory for the selection of the initiation site. In fact, they turn out to be not absolutely necessary for the initiation reaction: there exists non-SD and leaderless mRNAs in the cell that do not contain any SD sequence and yet they are translated efficiently. Additionally, the 30S ribosomal subunits, which were reconstituted with 16S rRNA from which the anti-SD segment was deleted, functioned effectively in initiation [25].

The participation of the SD sequence in the initiation reaction, however, has been convincingly demonstrated. SD sequences were isolated base paired to the anti-SD segments of the 30S ribosomes from a reaction mixture in which ribosomes were incubated with mRNA and fMet-tRNA [26]. Although the SD interaction does participate in the initiation reaction, if present in the IS, it does not appear to be absolutely required, or obligatory. Thus, there is a conflict between the hypothesis and reality because the hypothesis postulates that the SD interaction is the only pathway for initiation, that is, it is obligatory. A resolution can be provided for this problem by the CS mechanism if it is accepted as the initiation mechanism for prokaryotes, which will be discussed briefly now and then later, more in depth.

According to the CS initiation mechanism of protein synthesis the binding of the ribosome to the IS occurs one or a few subsites (sub-segments) at a time, but, except for the initiator codon containing subsite, none of the subsite interactions is absolutely essential as long as the IS is anchored to the ribosome by a sufficient number of adjacent subside interactions. This would be the case with the SD interaction if it is considered as one of the multiple subsite interactions of the CS mechanism, it would not always be essential. In other words, initiation of proteins synthesis is not solely dependent on the SD interaction and it is not obligatory.

3.2. Scanning hypothesis for eukaryotes

Since there were no obvious initiation recognition signals in eukaryotic mRNAs, it was proposed that the ribosomes do not outright recognize the IS, but only the 5′ end of the mRNA [2]. According to this proposal, the process begins with the 40S ribosome-Met-tRNA complex, which first recognizes and binds to the 5′ end of the mRNA [2], found to consist of a 7-methyl guanosine and referred to as cap [27]. The ribosomal complex then scans the initiation site for the first AUG codon, which was found subsequently not always to be the initiator codon so modifications were made to account for the findings [28]. Later studies have shown that initiation factors (eIF-4F) facilitated the binding of the 40S ribosomal subunit to the mRNA [29]. When the initiator codon is located, the 60S ribosome joins the 40S complex to form the 80S ribosome complex, an aminoacyl-tRNA is bound next and the first peptide bond is formed.

Numerous exceptions to the mechanism have been observed where eukaryotic ribosomes do bind directly to internal sites of the mRNA and initiate synthesis. The observations were construed as evidence for another, separate pathway, referred to as the cap-independent or IRES (internal ribosome entry sites) –mediated translation [5]. According to advocates of the IRES mechanism, a complex IRES RNA structure of the initiation site somehow promotes the correct binding of the 40S ribosomes to internal sites. If as concluded earlier, however, that the prokaryotic initiation signals, along with the prokaryotic initiation mechanism, are conserved in the eukaryotic ribosomal system is true, then there is a simple and alternative explanation for the IRES observations than a separate IRES synthetic pathway.

Before turning to the alternative explanation for the IRES observations, the weaknesses of the proposals of the scanning and the IRES reactions will be addressed. The proposed two mechanistically quite different reactions immediately pose a difficult problem: the same ribosome cannot perform the two very different functions, even with the aid of auxiliary proteins. This, implausibly, would require the evolution of another species of ribosomes. A related major weakness is the absence of a gradual evolutionary change in the scanning proposal: there would not be a gradual evolutionary change in the case of a prokaryotic ribosome that recognizes the IS, binds directly to it, changes to a eukaryotic ribosome that recognizes only the end of the mRNA, and then scans the IS for the initiator codon. Such a process would involve too great a change in the mechanistic characteristics of the ribosome. These shortcomings of the scanning mechanism suggested the proposal of the modified CS initiation mechanism for eukaryotes, which will be reviewed below.

The above complications may be resolved if the modified CS mechanism for the initiation of eukaryotic protein synthesis is indeed adopted as the mechanism for eukaryotes. This mechanism postulates its evolution from the prokaryotic mechanism [6,10,11]. According to this proposal, the initiation of eukaryotic protein synthesis involves two steps. The first step now needs to be restated in less specific, but broader terms. Instead of specifying that an initiation factor complex (eIF-4F) binds the cap of the mRNA, it is now proposed that evolved initiation factors, including eIF-4F, bind to, or interact with the mRNA [30]. Thus, the revised version of step 1 is: the evolved initiation factors, including eIF-4F, first bind to, or interact with, the mRNA. These bindings or interactions are assumed to make the IS eminently accessible, and enhance step 2, which is the initiation of translation by the conserved prokaryotic CS mechanism. As mentioned before, in the CS mechanism the ribosome binds directly to the IS of the mRNA, without ribosomal scanning of the IS. In summary, this modification of the prokaryotic CS initiation mechanism only adds the participation of evolved initiation factors to the function of the basic prokaryotic CS mechanism, and thus renders the eukaryotic model compatible with the proposal of a universal initiation mechanism, basically identical in all domains of life.

The alternative explanation for the IRES observations can now be conveniently discussed. From the viewpoint of the modified CS initiation mechanism, the existence of the scanning mechanism is only theoretical. There is no experimental evidence that proves that this mechanism actually operates in eukaryotic protein synthesis. Furthermore, the existence of the IRES pathway as a separate protein synthesizing pathway is questioned. The IRES pathway is viewed only as the *in vitro* expression of the activity of step 2 of the modified CS initiation mechanism without the expression of step 1. The IRES pathway is, in other words, only the *in vitro* expression of the conserved prokaryotic initiation mechanism in the eukaryotic ribosomes. This, then, is the alternative explanation for the IRES observations.

3.3. Cumulative specificity hypothesis for prokaryotes and eukaryotes

As mentioned earlier, the CS initiation mechanism of protein synthesis has its origin in a 1966 proposal that provided an essentially unique accessibility of the IS mechanism for prokaryotes [3]. It postulated that the initiator codon is selected by virtue of its unique accessibility. All non-initiator internal methionine codons were assumed to be sequestered and inaccessible to the ribosomes by secondary structure. This assumption was based on the observation that synthetic polynucleotides containing all four bases in equal proportions in random sequence, failed to act as mRNA in a cell-free protein synthesizing system, which was interpreted as that all AUG codons in the synthetic polynucleotide were inaccessible to ribosomes because of secondary structure (unpublished experiment). An experiment much later, however, indicated the need for an extension of the proposal. In that study, a non-SD model mRNAs was prepared for kinetic measurements [31]. The mRNAs—created to minimize secondary\structures —had an accessible AUG but no other obvious recognition signal, and yet, they were able to able to direct the initiation of polypeptide synthesis, apparently in full agreement with our proposed mechanism. However, a second, still unrestrained and thus supposedly accessible AUG, failed to act as an initiator. This seems to indicate that that the ribosomes were somehow able to discriminate the IS negatively by rejecting certain bases surrounding the initiator codon.

The unique accessibility proposal was thus modified to postulate that a site containing a non-initiator methionine codon can be made functionally inaccessible by sequestration with secondary structure or by other unfavorable local interactions, i.e., through steric hindrance, hydrophilic/hydrophobic mismatch, or by electrostatic repulsion, which all contribute negative specificity. The proposal was renamed as the unique accessibility hypothesis. However, when more was learned about the novel features of the CS model reaction for HIV proteases [9], the model reaction was incorporated into the unique accessibility proposal. The incorporation of CS reaction replaced the discrimination of ISs by the negative specificity of the unique accessibility hypothesis with the positive recognition of the IS by cumulative specificity. The modified reaction was renamed as the CS hypothesis for the initiation of protein synthesis [4].

The CS initiation mechanism of protein synthesis incorporates the key features of the CS model reaction of HIV proteases and postulates that recognition of the initiation signal is the result of interactions of one or a few subsites (sub-segments), at a time, between the ribosome and the IS of the mRNA. Thus, the selection of the IS occurs through cooperativity and cumulative specificity of subsite interactions that allow a reaction to occur even if not all subsites are occupied [4,32]. This enables many subsites of the IS that share only some of the structural elements to be accepted as ligands by the ribosomal subsites, and hence, the broad substrate specificity of the protein synthesizing system follows.

According to the CS model for the HIV protease reaction, none of the individual subsite interactions of the substrate active site with the respective subsite of the enzyme is absolutely essential as long as the peptide is properly anchored at a sufficient number of adjacent subsites. The same rule applies to the CS initiation of protein synthesis that none of the subsite interactions is absolutely essential except that the interaction of the initiator codon subsite of the IS and the ribosomal subsite with the accessible Met-tRNA anti-AUG codon is absolutely required. In the case of the initiation of protein synthesis, there must also be a sufficient number of adjacent subsite interactions to anchor the IS properly to the ribosome.

Another important feature of CS initiation mechanism is the role played by the secondary structure of the mRNA. It keeps the ISs accessible to the ribosomes and it also reduces the accessibility of non-initiator methionine codons by sequestering them and thus favoring the recognition of the initiator codon. The multiple roles played by the secondary structure were demonstrated in a study in which the secondary structure of bacteriophage f2 RNA was disrupted by treatment with formaldehyde. The treated RNA was shown to yield three new, non-viral polypeptides when the RNA was translated in an E. coli cell-free protein synthesizing system [33]. Three new non-initiator methionine codons were evidently selected as initiators because filtering by secondary structure was eliminated, which showed that the specificity of the ribosome alone was not sufficient to eliminate all of the non-initiator methionine codons.

Cumulative specificity in-reading-frame binding of ribosomes to mRNA

The proposed CS initiation mechanism [6,10, 11] for the in-reading-frame binding of the mRNA by the ribosomes will now be reviewed, critically evaluated as were the SD and scanning hypotheses, and then revised with an admission of mea culpa by one of the authors (T.N.). This revision of the in-reading-frame binding aspect of the basic CS initiation mechanism must not be confused with the previously described modified CS initiation mechanism, which only added the participation of evolved initiation factors to the function of the basic prokaryotic CS mechanism for the initiation of eukaryotic protein synthesis.

The initiation reaction was postulated to begin with a relatively strong interaction of the small ribosomal subunit initiation complex with an accessible subsite of the IS that contains the initiator AUG codon. The base pairing of the initiator AUG codon and the anti-AUG codon of the ribosome bound initiator Met-96tRNA, along with the strong binding of the entire sub-segment, secures the mRNA onto the ribosomal complex in reading frame. This first interaction is stabilized by subsite interactions that reach out in both directions of the AUG subsite in zipper-like fashion, until the ribosomal complex is firmly bound to the mRNA. Initiation can then occur when the first designated aminoacyl-tRNA is bound.

According to the above proposal, the ribosome binds to the mRNA in reading frame at the initiator codon containing subsite in a first interaction with the mRNA. This remarkable feat is possible only because of a probably invalid assumption of the proposal, namely that the initiator codon of the IS of about 50 nucleotides is always accessible for binding to the ribosomal initiation complex. As mentioned earlier, for the 30S ribosomal subunit, as an example, to bind, meshed perfectly, to a nucleotide sequence of about 47 nucleotides long, which is the length of the model E. coli IS, is no easy feat. This is because RNA interacts intra-molecularly so extensively by base pairing. A synthetic RNA polymer containing 4 bases in equal proportions in random sequence has been found to have about 50% of its bases paired [21]. Therefore, the mRNA would not likely to have its initiator codon located in the middle of the IS of 47 nucleotides more accessible to the ribosome than any other IS subsites, even in natural mRNAs. Prokaryotic viral mRNAs have been shown to have 60-70% of their nucleotides involved in base pairing [34]. However, one cannot rule out the possibility that secondary structure increased with a precise arrangement could enhance the accessibility of the initiator codon subsite. So the possibility of an intrinsically accessible initiator codon subsite does exist.

The major problem in binding of the mRNA IS by the ribosome is that the IS, which is about 50 nucleotides long, must minimize not only its base interactions within the IS, but also its intra-molecular interactions with nearby adjoining regions, and even with distant regions of the mRNA. Local interactions may be minimized by appropriate evolutionary base selections, that is, by controlling the primary sequence of the IS. There are studies indicating that regions around and at initiation sites are low in secondary structure. One group of researchers determined the secondary structure of the region by computer analysis of the nucleotide sequence of the intracistronic initiation sites of infB mRNA for various bacterial species [35]. They found that the mRNA has an open structure around the initiation site. A second group computationally folded human and mouse mRNA sequences on sets of transcripts, and found that the initiation site is characterized by a relaxed secondary structure [36].

In any case, a significant portion of the IS, nonetheless, must be accessible to the ribosome, for otherwise, one would not have a functioning mRNA. The problem, then is surmised to be, that despite the accessibility of the IS to the ribosome, the IS may not be fully extended with the initiator codon subsite readily accessible. Therefore, some mechanism is needed to stretch or to extend the IS on the ribosome so that the initiator codon subsite is accessible. As will be suggested by the revised basic CS initiation mechanism below, this can be done by anchoring the IS onto the ribosome at a sufficient number of adjacent subsite interactions. Such a mechanism essentially stretches the IS to make the initiator codon subsite accessible to the appropriate ribosomal binding subsite.

The original proposal, that is, the prokaryotic CS initiation mechanism, will now be revised and recapitulated. The revised proposal, which assumes that all subsites of the IS, are more or less, equally accessible, postulates that the first subsite interaction will occur, more or less randomly, between one or more of all of the various subsites of the ribosome and its or their respective IS subsites. If the first interaction happens to be between the ribosomal subsite with the anti-AUG codon of the initiator Met-tRNA and the initiator AUG codon containing IS subsite, then the remaining subsites will interact in zipper-like fashion to stabilize the first interaction. In this manner an in-reading-frame binding of the mRNA by the ribosome will be completed. In all other first subsite interactions, the critical interaction which involves the anti-AUG codon of the initiator Met-tRNA and the initiator codon IS subsite will occur only after the IS is anchored onto the ribosome by a sufficient number of adjacent IS subsites and the initiator codon subsite is made accessible. In other words, the critical subsite interaction of the ribosomal subsite with the anti-AUG codon of the initiator Met-tRNA and the initiator codon IS subsite, occurs towards the end of the subsite interactions for all first non-initiator codon subsite interactions.

If, however, the IS initiator codon subsite is somehow intrinsically accessible because of the organized secondary structures of the mRNA, then all pathways will be a single pathway as originally proposed, i.e., al first ribosomal subsite bindings will be of the initiator codon IS subsite, and the remaining subsites will interact in zipper-like fashion to stabilize the first interaction. The in-reading-frame binding of the mRNA by the ribosome will all be completed in this manner. Another way in which the preceding single pathway may predominate under conditions in which the initiator codon subsite is not intrinsically accessible and the ribosome interacts with it only randomly, is the favorability or strength of the first interaction as described below.

Evidence for the strength of the critical subsite interaction where the initiator codon containing IS subsite is bound to the ribosomal subsite with the anti-AUG codon of the initiator Met-tRNA is the strength of the specific base pairing of the initiator AUG codon and the anti-AUG codon of the initiator Met-tRNA. Further evidence for the strength of this interaction was provided by studies in which nucleotides around the AUG initiation codon were replaced. Protein synthesis was decreased by as much as 95% upon replacement of three nucleotides adjoining the initiator AUG codon on the 5′ side [37]. Similarly, the replacement of three nucleotides just next to the AUG codon on the 3′ side decreased translation by more than 65% [38]. These observations underscore the importance of the AUG segment in initiation, and suggest a strong interaction of the ribosome with this particular IS subsite.

The process of the binding of ribosomes to leaderless mRNA is as described above for canonical mRNAs, except that the binding of eIF4F may not occur and subsite interactions are only between the ribosomal binding subsites and the initiator codon subsite and the amino acid coding region subsites of the IS . The final step in the initiation of translation of canonical and leaderless mRNAs is the binding of the aminoacyl-tRNA directed by the codon following the initiator codon, and the formation of the first peptide bond.

Base pairing is probably not the only means of molecular recognition of the nucleotides of the IS by the ribosome in the interactions, since the ribosomal binding site is composed of RNA and proteins. In the interaction of the ribosomal subsites containing the anti-SD sequence, base pairing is the predominant means of nucleotide recognition when the SD sequence is present in the mRNA. The Shine-Dalgarno base-pairing interaction may be considered as just one of the multiple independent interactions of the CS initiation mechanism. Recognition in other ribosomal binding sub-sites may also involve steric fit, steric hindrance, hydrophilic or hydrophobic match or mismatch, and electrostatic attraction or repulsion. In other words, the recognition is also the product of both positive and negative specificities.

The revised CS hypothesis for the initiation of protein synthesis assumes that the productive positioning of the initiator codon of the IS on the ribosome results from the cumulative effect of independent interactions between each base and its respective subsite on the ribosome. According to this view, except for the subsite containing the initiator codon, none of these individual interactions is absolutely essential as long as the IS is properly anchored at a sufficient number of adjacent subsites. It is assumed that, to make the initiator codon region accessible, enough adjacent subsite interactions are needed to anchor the IS to the ribosome, which extends or stretches out the IS. The extension or stretching of the IS exposes the initiator codon region, making the initiator codon accessible. This allows the in-reading-frame binding of the ribosome to the mRNA to be completed by the interaction of the ribosomal subsite with the accessible anti-codon of the initiator Met-tRNA and the IS subsite containing the initiator codon AUG. In this manner the base-pairing of the AUG and the anti-AUG codon of the ribosome bound initiator Met-tRNA can occur.

Thus, it follows that recognition of the IS does occur by recognition of a number of individual subsites. As mentioned several times, it is most unlikely that collision of the ribosome binding site with the IS of the mRNA would be a single step in which the ribosome binding site and the IS would already be perfectly oriented to achieve an optimal fitting of all subsites. Rather, the initial collision probably results in the binding at just one or a few of the subsites, each contributing incrementally to anchoring the IS to the ribosome. Then if these interactions are favorable, the rest of the subsites will be filled in cooperatively, in a zipper-like fashion. The ultimate result is, however, the binding of the whole IS as a block, and the exact positioning on the ribosome of the subsite containing the initiator codon. In that ultimate ribosome-mRNA complex the global strength of binding and the precise positioning of the initiator codon subsite of the IS at the reaction site still depends on the sum of the contributions of the individual subsites, i.e., the cooperativity of the subsites.

4. Conclusion

It is important to understand that this evaluation of current hypotheses have the advantage of hindsight provided by knowledge of the initiation of protein synthesis, not available when the SD and the scanning hypotheses were proposed. Critical evaluations are thus made while appreciating the great value of the two older hypotheses in stimulating research.

Returning to the conclusion of this review, we have essentially taken the view that there are three keys to unlocking the secrets of the initiation of protein synthesis. The first two keys provide insights into the nature of the initiation mechanism, and the third key is the initiation mechanism that is compatible with those insights.

The first key consists of the implications of the characteristics of the IS. The E. coli model IS consists of a nucleotide sequence of about 47 bases, with preferred bases in given positions, but no particular base in any given position, except for the initiator codon located in a specific position in every IS of the mRNA. An important feature of the model IS is that all, or nearly all of its nucleotides presumably have signal character, which includes the leader and amino acid coding regions. This means that the initiation signal of a comprehensive mechanism must include the leader, as well as, the amino acid coding regions. The length of the model IS indicates that the IS is extensive, nonrigid, and complex, and that the ribosome is unlikely to bind the IS of the mRNA, meshed perfectly, in a single collision with the mRNA. The unlikelihood of the binding of the mRNA by the ribosome in a single collision predicts that the ribosome would bind the IS of the mRNA via a sub-segment or a few sub-segments, at a time. The ribosomal binding of the IS of the mRNA thus must happen between the ribosomal subsites and the respective IS subsites, one or a few at a time. The initiation mechanism must also account for the recognition of initiation signals consisting of about 47 nucleotides, at least in E coli, with preferred bases in given positions.

The second key consists of the implications of evolutionary evidence and logic that point to a universal initiation mechanism in all domains of life. This implies that a common or universal initiation mechanism should be constantly favored, even when one is faced with an appealing mechanism for a particular phylum, and when one examines any data, always being alert for indications of conservation of initiation signals or mechanisms. As reviewed earlier in this chapter, when such diligence was maintained in reviewing the studies of heterologous systems in which ribosomes or mRNAs of different phyla were interchanged, the conclusion was reached that prokaryotic initiation signals as well as the underlying prokaryotic initiation mechanism are conserved in eukaryotes. The conservation of the initiation signals and the initiation mechanism has been interpreted as evidence of the existence of a universal initiation mechanism.

The third key consists of the initiation mechanism for protein synthesis that is most compatible with the insights provided for a mechanism by keys 1 and 2. Unfortunately the SD and scanning hypotheses were proposed before publication of the study revealing the IS characteristics of E. coli, although many base sequences of IS were known. That evolutionary evidence and logic supported a universal initiation mechanism was not unknown, but the proposal of an initiation mechanism for eukaryotes vastly different for that of prokaryotes indicated a lack of conviction in a universal mechanism, or at least, in the evidence for it. Thus, key 1 was not in existence at the time of proposal of the two hypotheses, and there was not much faith in key 2. For these reasons, each of the SD and the scanning hypotheses has only considered its own particular facet of the mechanism and ignored most of the insights provided by keys 1 and 2.

For example, despite the observation that the signal character of the IS is divided about half in the leader region and the other half in the amino acid coding region, the SD hypothesis postulates an initiation signal located exclusively in the leader region, less than 20% of the nucleotides with signal character, while the scanning hypothesis postulates a single nucleotide at the 5′ end of the mRNA as a recognition signal, not even included in the IS. For this reason, the two hypotheses cannot account for the initiation of translation of leaderless mRNAs. Furthermore, the two hypotheses also do not acknowledge the problem of the need of ribosomes to bind to an extensive, nonrigid and complex IS, sub-segment by sub-segment, nor do they address the possibility of a universal initiation mechanism. The SD and the scanning hypotheses, in other words, essentially ignored most of the insights provided by keys 1 and 2.

Recapitulating, in the view described above, the SD as well as the scanning hypotheses hardly account for the insights of keys 1 and 2, i.e., the implications of the characteristics of the IS of the mRNA and of the dictum of universality of the initiation mechanism. The CS hypothesis, on the other hand, is more compatible with the insights provided by keys 1 and 2. The CS mechanism for initiation of prokaryotic synthesis is essentially the mechanism formulated for the HIV proteases with minor changes to adapt it to the initiation of protein synthesis. This mechanism postulates a cooperative and cumulative, sub-segment by sub-segment recognition binding of the IS. As initiation signals, the CS mechanism recognizes nucleotide sequences with preferred bases in given positions, that is, the entire IS with signal character.

Finally, the CS initiation mechanism for prokaryotes modified for eukaryotic protein synthesis appears to be in accord with all experimental observations. The proposal postulates an evolutionary link of the initiation mechanism of eukaryotic protein synthesis to that of the prokaryotes. It assumes that evolved eukaryotic initiation factors interact with the mRNA and make the IS eminently accessible. This dramatically enhances the ribosomal binding to the IS and greatly increases the rate of initiation of protein synthesis by the conserved CS mechanism in the eukaryotic ribosomes. This modification keeps the eukaryotic initiation mechanism basically identical to the prokaryotic mechanism, and therefore, one may conclude that the two mechanisms are essentially identical as well as universal. The modified CS mechanism is compatible with the conclusion that the prokaryotic initiation signals and the prokaryotic initiation mechanism are conserved in eukaryotes.

Author details

Tokumasa Nakamoto and Ferenc J. Kezdy
Department of Biochemistry and Molecular Biology, The University of Chicago, Chicago, IL, USA

Acknowledgement

The authors offer their profound thanks to Dr. Herbert Friedmann, a colleague at The University of Chicago, for his conscientious and excellent editing in the preparation of this manuscript.

5. References

[1] Shine J, Dalgarno L (1974) The 3'-terminal sequence of Escherichia coli 16S ribosomal RNA: complementary to nonsense triplets and ribosome binding sites, Proc. Natl. Acad. Sci. USA, 71: 1342-1346.

[2] Kozak M (1978) How do eukaryotic ribosomes select initiation in regions in messenger RNA?, Cell, 15: 1109-1123.

[3] Kolakofsky D, Nakamoto T (1966) The initiation of viral protein synthesis in E. coli. Extracts, Proc. Natl. Acad. Sci. U.S.A. 56: 1786–1793.

[4] Nakamoto T (2007) The initiation of eukaryotic and prokaryotic protein synthesis: A selective accessibility and multisubstrate enzyme system, Gene, 403: 1-5.

[5] Hellen CUT, Sarnow P (2001) Internal ribosome entry sites in eukaryotic mRNA molecules, Genes Dev., 15: 1593-1612.

[6] Nakamoto T (2009) Evolution and the universality of the mechanism of initiation of protein synthesis, Gene, 432: 1-6.

[7] Scherer GFE, Walkinshaw MD, Arnott S, Morre SDJ (1980) The ribosome binding sites recognized by E. coli ribosomes have regions with signal character in both leader and protein coding segment, Nucl. Acid Res., 8: 3895–3907.

[8] Kozak M (1984) Compilation and analysis of sequences upstream from the translational start site in eukaryotic mRNAs, Nucl. Acid Res., 12: 857-872.

[9] Poorman RA, Tomasseli AG, Heinrikson RL, Kézdy FJ 1991) A cumulative specificity model for proteases from human immunodeficiency virus types 1 and 2, inferred from statistical analysis of an extended substrate database, J. Biol. Chem. 266: 14554-14561.

[10] Nakamoto T (2010) Mechanisms of the Initiation of Protein Synthesis: In Reading Frame Binding of Ribosomes to mRNA, Mol. Biol. Rep. DOI 10.1007/s11033-010-0176-1

[11] Nakamoto T (2011) Mechanisms of the Initiation of Protein Synthesis: In Reading Frame Binding of Ribosomes to mRNA, Mol. Biol. Rep. 38:847–855

[12] Rekosh DM, Lodish HF, Baltimore D (1970) Protein synthesis in Escherichia coli extracts programmed by poliovirus RNA, J. Mol. Biol. 54: 327-340.

[13] Glover JF, Wilson MA (1982) Efficient translation of the coat protein cistron of tobacco mosaic virus in a cell-free system from *Escherichia coli*, Eur. J. Biochem., 122: 485–492.

[14] Siegert W, Konings RNH, Bauer H, Hofschneider PH (1972) Translation of avian myeloblastosis virus RNA in a cell-free Lysate *of Eschericihia coli*, Proc. Natl. Acad. Sci. USA, 69: 888-891.

[15] Morrison TG, Lodish HF (1973) Translation of bacteriophage Qβ RNA by cytoplasmic extracts of mammalian cells, Proc. Natl. Acad. Sci. USA, 70: 315–319.

[16] Wang S, Marcu KB, Inouye M (1976) Translation of a specific mRNA from *Escherichia coli* in a eukaryotic cell-free system, Biochem. Biophys. Res. Comm., 68: 1194-1200.
[17] Paterson BM, Rosenberg M (1979) Efficient translation of prokaryotic mRNAs in a eukaryotic cell-free system requires addition of a cap structure, Nature, 279: 692-696.
[18] Rosenberg M, Paterson BM (1979) Efficient cap-dependent translation of polycistronic prokaryotic mRNAs is restricted to the first gene in the operon, Nature 279: 696-701.
[19] Nomura M, Lowry CV (1967) Phage f2 RNA-=Directed Binding of Formylmethionyl-tRNA to Ribosomes and the Role of 30S ribosomal Subunits in Initiation of Protein Synthesis, Proc. Natl. Acad. Sci. USA 58: 946-953.
[20] Gualerzi CO, Pon CL (1990) Initiation of mRNA Translation in Prokaryotes. Biochem 29: 5881–5889.
[21] Gralla J, Delisi C (1974) mRNA is Expected to Form Stable Secondary Structures. Nature 248: 330–332.
[22] Merrick WC (2004) Cap-independent and cap-dependent translation in eukaryotic systems, Gene 332: 1-11.
[23] Nakamoto T (2006) A unified view of the initiation of protein synthesis, Biochem. Biophys. Res. Commun. 341: 675–678.
[24] Steitz J.A. (1979) Genetic signals and nucleotide sequences in messenger RNA,in Biological Regulation and Development (Goldberger, R. ed.) pp 349-399, Plenum Publ. Co., New York.
[25] Melancon P, Leclerc D, Destroismaisons N, Brakier-Gingras L (1990) The anti-Shine–Dalgarno region in Escherichia coli 16S ribosomal RNA is not essential for the correct selection of translational starts, Biochem., 29: 3402–3407.
[26] Steitz JA, Jakes K (1975) How ribosomes select initiator regions in mRNA: base-pair formation between the 30 terminus of 16S rRNA and the mRNA during initiation of protein synthesis in *Escherichia coli*, Proc. Natl. Acad. Sci. USA 72: 4734–4738.
[27] Shatkin J (1976) Capping of eukaryotic mRNAs, *Cell* 9: 645-653.
[28] Kozak M (1995) Adherence to the first-AUG rule when a second AUG codon follows closely upon the first, Proc. Natl. Acad. Sci. USA, 92: 2662-2666.
[29] R.D. Abramson, T.E. Dever, T.G. Lawson, B.K. Ray, R.E. Thach, W.E. Merrick, The ATP-dependent interaction of eukaryotic initiation factors with mRNA, J. Biol. Chem., 262 (1987) 3826-3832.
[30] Anthony DD, Merrick WC (1991) Eukaryotic initiation factor (eIF)-4F: Implications for a role in internal initiation of translation, J. Biol. Chem., 266: 10218-10226.
[31] Calogero RA, Pon CL, Canonaco A, Gualerzi CO (1988) Selection of the mRNA translational initiation region by Escherichia coli ribosomes, Proc. Natl. Acad. Sci. USA, 85: 6427-6431.
[32] D.E. Koshland Jr DE, Hamadani K (2002) Proteomics and models for enzyme cooperativity, J. Biol. Chem., 27: 46841-46844
[33] Lodish, HF (1970) Secondary structure of bacteriophage f2 ribonucleic acid and the initiation of in vitro protein biosynthesis. J. Mol. Biol. 50: 689-702.
[34] Ricard B, Salser W (1976) Optical measurements reveal base-pairing inT4-Specific mRNAs, Biochim. Biophys. Acta 425: 196–201.

[35] Laursen BS, de A. Steffensen SA, Hedegaard J, Moreno JMP, Mortensen KK, Sperling-Petersen HU (2002) Structural requirements of the mRNA for intracistronic translation initiation of the enterobacterial *infB* gene, Genes to Cells, 7: 901-910.

[36] Shabalina SA, Ogurtsov AY, Nikolay A, Spiridonov NA (2006) A periodic pattern of mRNA secondary structure created by the genetic code, Nucl. Acid Res., 34: 2428-2437.

[37] Hui A, Hayflick J, Dinkelspiel K, de Boer HA (1984) Mutagenesis of the three bases preceding the start codon of the galactosidase mRNA and its effect on translation in *Escherichia coli*, EMBO J., 3: 623-629.

[38] Looman AC, Bodlaender J, Comstock J, Eaton D, Jhurani P, de Boer HA, van Knippenberg PH (1987) Influence of the codon following the AUG initiation codon on the expression of a modified *lac Z* gene in *Escherichi coli*, EMBO J., 6: 2489-2492.

Chapter 7

Protein Synthesis and the Stress Response

Assaf Katz and Omar Orellana

Additional information is available at the end of the chapter

http://dx.doi.org/10.5772/50311

1. Introduction

Accuracy of protein synthesis is critical for life since a high degree of fidelity of the translation of the genetic information is required to accomplish the needs of the cellular functions as well as to preserve the variability developed by evolution. More than one hundred macromolecules are involved in this process even in the simplest organisms, including ribosomal proteins, translation factors, aminoacyl-tRNA synthetases as well as ribosomal and transfer RNAs (being rRNAs near 80% of total cellular RNAs in bacteria). Accuracy of translation of the genetic information is accomplished at different levels, being one of the most relevant the specific interaction of the aminoacyl-tRNA synthetases with their substrates, tRNA and amino acids. Either specific molecular interactions that avoid the miss-incorporation of amino acids or the hydrolysis of wrong aminoacyl-tRNAs represent strategies utilized by the aminoacyl-tRNA synthetases to reduce the formation of miss-acilated tRNAs. Additionally, the discrimination by the elongation factor Tu (EF-Tu) against miss-acylated aminoacyl-tRNAs over correctly acylated increases the accuracy of incorporation of the proper aminoacyl-tRNAs to the ribosome. The accurate decodification of the mRNA by the incorporation of the aminoacyl-tRNA with the correct anticodon ensures the fidelity of translation of the genetic information. The consequence of these discriminatory events led to an accuracy of translation of the genetic information to as low as 10^{-4} errors per amino acid incorporated. However, certain level of inaccuracy might be tolerated, even might be beneficial to the cells under certain physiological and environmental conditions.

How does the cellular translational machinery respond against stress conditions? Is protein synthesis as accurate as in normal conditions when cells are stressed? These are topics that are not well understood. Different kinds of stressing conditions might have different responses in the components of the protein synthesis system. In this chapter, we will briefly describe how bacteria respond to two stress conditions, oxidative stress and amino acid starvation, that microorganisms are commonly exposed to including in environmental conditions.

2. Oxidation of the translation machinery during oxidative stress

Oxidative stress is defined as the condition where oxidative species production is faster than the ability of the cell to eliminate them and reduce or degrade oxidized products, leading to damage [1–3]. This condition can be met when production of or exposure to oxidative species is increased, such as when bacterial cells are attacked by the immune system of a host [4–7] or when oxygen reactive species (ROS) are generated in cyanobacteria as a byproduct of photosystem II irradiation with strong light [8]. Also, "normal" ROS levels can cause oxidative stress when there is a lower level of protective enzymes as when bacteria pass from an anaerobic environment to another with oxygen [9]. As a consequence of the increase in oxidative species concentration several macromolecules can be modified, including proteins, RNA, DNA and lipids [10–12]. In this chapter we are mainly interested in the oxidation of the first two, proteins and RNA, as they are directly involved in the translation process. Several amino acids and cofactors of proteins can be target of oxidation *in vivo*. Within these, the most sensitive to oxidation are Fe/S clusters, methionines and cysteines [11,12]. Some of the oxidation states of cysteine (sulfenic acid and disulfide bond) can be reduced so this amino acid is frequently involved in regulatory processes [13–15]. The rest of the amino acids are usually considered to be more stable to oxidation, although many of them are carbonylated during oxidative stress, especially when they are located near to cations that can catalyze the oxidation reaction [11,12,16,17]. RNA oxidation is less well studied, although we know that ribonucleotide bases can be oxidized to 5-hydroxyuridine, 5-hydroxycytidine, 8-hydroxyadenosine and 8-hydroxyguanosine [18]. In addition to these oxidative modifications, nucleotides can also completely loss their bases during oxidative stress [19].

When cells enter in oxidative stress several targets are oxidized decreasing the activity of many metabolic pathways including translation in both bacteria [4,20–23] and eukaryotes [24–27]. Part of this inhibition might be due to the oxidative inactivation of several enzymes involved in the metabolism of energy and amino acid synthesis [7,11,12,28,29] both of which are essential for protein synthesis. Also, several macromolecules that participate in translation have been found to be a target of oxidation in *in vivo* or *in vitro* experiments indicating that translation is directly targeted by oxidative species. In bacteria, these target macromolecules include elongation factors Tu [30–34], Ts (EF-Ts) [32] and G (EF-G) [8,9,22,32,35], several ribosomal proteins [31,36,37], tRNA [38–42] and aminoacyl-tRNA synthetases (aaRS) [31,34,36,37,43,44] (Table 1). rRNA and mRNA has also been shown to be oxidized *in vivo* in eukaryotes [19,25,26], but in bacteria this has not been tested.

Although many macromolecules involved in translation have been found to be target of oxidation, there is little information on the effects of this oxidation on translation and the bacterial physiology (Figure 1). The effects of oxidation of any of the ribosomal proteins or of elongation factors Tu and Ts have not been studied, although we know that the deletion of yajL (that encodes for a chaperon/oxido-reductase that protects this and other proteins during oxidative stress) increases the error rates of translation when *Escherichia coli* is incubated with a moderate concentration of H_2O_2 (~100 μM)[34,45]. Conversely, the effect of oxidation of EF-G has been characterized in a much greater detail. In *E. coli* cells, EF-G has been shown to be carbonylated in several stress conditions including incubation with H_2O_2, menadione or paraquat, transference from anaerobic to aerobic conditions in high iron concentrations [9] and growth arrest conditions in strains deleted for superoxide dismutase [35]. EF-G has also been shown to be carbonylated in *Bacillus subtilis* cells growing exponentially or exposed to H_2O_2 [32]. Finally, in the cyanobacteria *Synechocystis* sp., oxidation of three isoforms of EF-G during oxidative stress caused by excessive light exposure has been shown to inhibit translation [8,20,21]

Target	Type	Reference
Ribosomal proteins		
Ribosomal proteins L5, L14, L7/L12, L27, L31, S2, S4, S17 and S21	Cysteine oxidation	[31,36,37]
Ribosomal proteins S1, S2, S3, S4, S8, S10, S11, S12, S19, L2, L5, L6, L10, L11, L12, L13, L14, L27, L28	Covalent binding to chaperon/oxido-reductase protein through cysteine bond	[34]
Ribosomal proteins S2, S4, S7, S11, S13, S18, L16, L17	Disulfide bond formation	[37]
Elongator factors		
EF-Tu	Cysteine oxidation	[30,31]
EF-Tu	Covalent binding to chaperon/oxido-reductase protein through cysteine bond	[34]
EF-Tu	Carbonylation	[32,33]
EF-Ts	Carbonylation	[32]
EF-G	Cysteine oxidation	[8,20,21,46]
EF-G	Carbonylation	[9,32,35]
Aminoacyl-tRNA synthetases		
Alanyl, Phenylalanyl, Glutamyl, Glycyl, Aspartyl, Leucyl, Isoleucyl, Seryl and Threonyl-tRNA synthetase	Cysteine oxidation	[31,36,37,43,44]
Alanyl, Isoleucyl, Leucyl, Threonyl and Phenylalanyl tRNA synthetases	Covalent binding to chaperon/oxido-reductase protein through cysteine bond	[34]
tRNA		
tRNA	4-thiouridine oxidation	[41,47]
tRNA	Cross-linking of 4-thiouridine in position 8 with C8.	[42,48,49]

Table 1. Some macromolecules involved in translation that have been shown to be oxidized in bacteria.

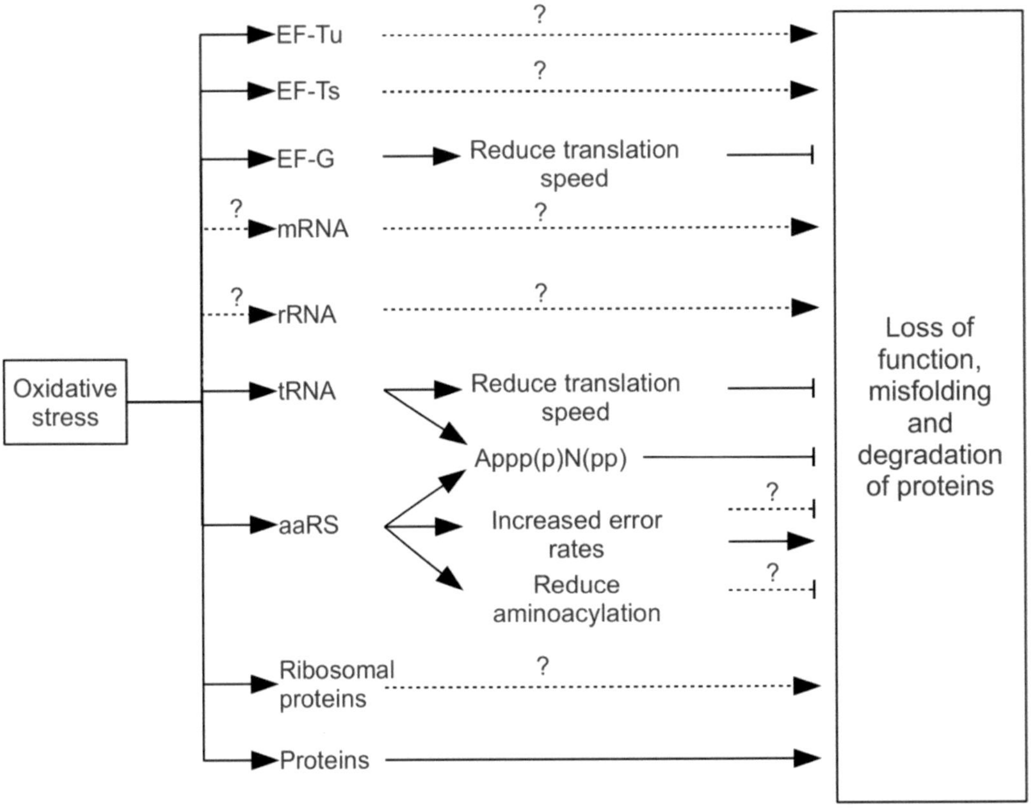

Figure 1. Effect on protein synthesis of the oxidation of translation related macromolecules. Oxidation of proteins may produce loss of function and structure which further produce protein aggregates that are toxic to the cell. Control of the translation machinery may sometimes protect the cell by decreasing the protein synthesis rate or enhancing the translation of specific proteins that protect the cell such as chaperons and proteases. Nevertheless, oxidation of the translation machinery may also increase its error rates which would produce proteins that are less stable increasing the toxicity of the oxidative stress.

2.1. EF-G oxidation in *Synechocystis* sp. during exposure to high intensity light

In *Synechocystis* sp., as well as in all photosynthetic organisms, light provides the energy that drives photosynthesis. Nevertheless, using photosynthesis is risky because photosynthetic transport of electrons or transfer of excitation energy to oxygen can produce ROS and subsequent oxidative stress [50]. Absorption of excessive light can also decrease the photosynthetic capacity of an organism in a process referred to as photodamage. The main target of photodamage is protein D1 which is part of photosystem II where the photochemical reaction and subsequent transport of electrons from water to plastoquinone occurs [51]. Loss of activity depends on the balance between both the speed of damage and repair of protein D1. The limiting step of repair is the *de novo* translation of the protein, thus the use of protein synthesis inhibitors like chloramphenicol has allowed to study damage separated from repair. These experiments have shown that, although oxidative stress increases the sensitivity of photosystem II to photodamage, its role is mainly to inhibit the translation of protein D1 while damage is caused directly by light [8,20,21]. [^{35}S]-Methionine incorporation experiments confirmed that D1 protein translation is inhibited under oxidative stress and also showed that this is a rather general effect where translation of all thylakoid membrane proteins is inhibited [20,21].

Most of D1 protein translation elongation is performed in membrane bound polysomes. During oxidative stress the fraction of mRNA associated to those polysomes decrease suggesting that the inhibition of translation happens at the elongation step [8,20,21]. Further *in vitro* experiments showed that the main target of oxidation is EF-G. *Synechocystis* sp genome encodes for 3 EF-G proteins, all of which can enhance protein translation in cell extracts that have been oxidized *in vitro* with H_2O_2 [22]. Mutation analysis of one of this EF-G variant, encoded by gene slr1463 of *Synechocystis* sp. PCC6803, showed that Cys105, which is near to the GTP-binding site, is the primary target of oxidation. Mutations on this cysteine or Cys242 (with which Cys105 form a disulfide bond) bears an EF-G protein that is resistant to H_2O_2 oxidation *in vitro* as measured by its ability to stimulate translation in oxidized cell extracts [8,46]. It has not been possible to substitute the slr1463 gene by a Cys105Ser mutant. Instead the mutant protein has been expressed in a strain of *Synechocystis* sp. that maintains its copy of the wild type gene. In this system, expression of slr1463 Cys105Ser decreased the loss of photosystem II activity and increased the rate of D1 translation confirming the relevance of this cysteine during oxidative stress inhibition of translation [52].

EF-G is substrate of reduction by thioredoxin in *Synechocystis* [46,53]. Based on this observation, it has been proposed that the translation activity could be rapidly recovered after oxidative stress through the reduction of cysteines 105 and 242 by thioredoxin. This last protein in turn would be reduced by NADPH-thioredoxin reductase and ferredoxin-thioredoxin reductase with reducing power from photosystem I [8,46].

2.2. Oxidation of tRNA

Together with the oxidation of ribosomal proteins and translation factors, other macromolecules involved in translation can be oxidized during oxidative stress. In this context, oxidation of RNA has been found in eukaryotes during oxidative stress, senescence or some age related diseases [19,25,26]. This reaction would depend on the ability of RNA to bind Fe^{+2} which catalyzes its oxidation. As mRNA and especially the more abundant rRNA bind Fe^{+2} better than tRNAs, these molecules would be more prone to oxidize [26]. The effect of oxidation over some specific mRNAs was shown to decrease translation of those genes through either increasing ribosome stalling or decreasing translational speed [25]. Oxidation of eukaryotic ribosomes also decreases translation efficiency, although in this case it is not clear whether oxidation of rRNA or ribosomal proteins is responsible of the effect [26].

In bacteria oxidation of rRNA and mRNA has not been studied, but on the other hand tRNA is known to be subject of oxidation *in vitro* [38,47] and *in vivo* [41,42]. Many of the tRNA bases are modified after its transcription [54] and apparently those modifications are very relevant for the role that tRNA plays during oxidative stress. In *E. coli* some of these modifications have a protective effect. For example, methylation of A37 in $tRNA_1^{Val}$ increases survival after an incubation of 2 hrs in 5 mM H_2O_2 [55]. The authors of this study cite unpublished data that indicates that several other tRNA modifications would improve the fitness of *E. coli* in such oxidative conditions. It has also been shown that deletion of several tRNA modification enzymes affect survival of *E. coli* in a milder oxidative stress condition (0.5 mM H_2O_2) [56]. Despite these reports, it is not clear how tRNA modifications improve survival to oxidative stress. It has been suggested that tRNA modifications increase the efficiency of translation in this condition, which would be necessary to cope with the degradation of proteins due to loss of function and/or structure [55]. There are some reports in the literature that show that an increase in error rates during protein synthesis enhances protein oxidation. In theory, this would be due to an increased exposure of oxidation targets that are normally hidden in the interior of proteins [3,57]. Thus, it is also possible that the requirement of tRNA modification during oxidative stress is related with the need of decreasing error rates to hide possible oxidation targets.

Nevertheless, in the only well studied example, tRNA modification is required because it directly participates in initiating the response to oxidative stress. An important source of oxidative stress for bacteria in environments exposed to sun light is near UV irradiation (300-400 nm), which corresponds to the sun irradiation with highest energy that can cross the atmosphere [58]. Near UV irradiation of bacteria like *E. coli, Salmonella Thyphimurium* or *Enterobacter cloacae* produces growth arrest which depends mainly in the photochemical oxidation of the tRNA modified base 4-thiouridine (s^4U) present in position 8 of all tRNAs [39–42]. In some tRNAs that also have a C in position 13 (50% of bulk tRNA) an internal cross-linking reaction happens that produces 5-(4′-pyrimidin 2′-one) [39,40,59]. Some cross-linked tRNAs have been shown to be poor substrates for aminoacylation [40,60,61] and in some cases also for translation [60]. The accumulation of such deacylated tRNAs can trigger the stringent response (see below). Thus, a combination of the trigger of stringent response plus the decrease in tRNA aminoacylation inhibits protein synthesis after UV exposure [40,62,63].

In *E. cloacae*, a short period of exposure to UV light protects the cells from the growth arrest produced by a second exposition of 60 minutes. This effect coincides with a decrease in the content of s^4U modification in tRNA [48], which could be interpreted as the presence of a protective effect mediated by the loss of this tRNA modification. Concordantly, mutation of the genes involved in s^4U modification of tRNAs (*nuv* mutant) also protect *S. thyphimurium* cells from the lag in the growth curves produced by short exposures (15 min) to near UV light. Nevertheless, the lag in growth seems to have a protective effect in long term exposure. After 4 to 5 hours of exposure to near UV, the mutant cells died at a faster rate than wild type (over 10 fold difference of survival after 8 hrs.). Part of this increased sensitivity to UV exposure is due to the lack of stringent response, but s^4U modification is somehow also relevant in the induction of several heat shock and oxidative stress response proteins such as alkyl hydroperoxide reductase [41]. Thus, it seems that the loss of s^4U modification after exposure to UV could be associated with a lag in the tRNA turnover process and not with a long term protection process, although similar long term exposures have not been performed with *E. cloacae*. In accordance, tRNA sulfuretransferase activity, which is necessary for the synthesis of s^4U tRNA modification, is inhibited in *E. coli* extracts from cells exposed to UV irradiation [49].

2.3. 5′-adenylyl dinucleotides and oxidative stress

Beside its direct effect on stringent response and the induction of several oxidative stress response enzymes, cross-linking of s^4U also induce the synthesis of 5′-adenylyl dinucleotides of the general structure AppppN (adenosine-5′, 5′′-ribonucleotide tetraphosphate) and ApppN (adenosine-5′, 5′′-ribonucleotide triphosphate) [41]. These dinucleotides are synthesized by aminoacyl-tRNA synthetases during stressful conditions such as oxidative stress and heat shock [41,64–67]. During the synthesis of aminoacyl-tRNA, aaRS produce aminoacyl-adenylate (aa-aMP) as an intermediate of the reaction (reaction 1). In this intermediate molecule the amino acid (aa) is activated for its transference to tRNA (reaction 2) [68], but the adenylate is also activated and can be transferred to either ATP, ADP, ppGpp or other nucleotides in a side reaction that forms dinucleotides of the general structure AppppN, ApppN or ApppNpp [69–71] (reaction 3). This reaction can be catalyzed by several aaRS, but the most active are Phenylalanyl-tRNA synthetase (PheRS) and specially Lysyl-tRNA synthetase (LysRS) [64,69].

$$\text{ATP} + \text{aa} \rightarrow \text{aa-AMP} + \text{PPi} \tag{1}$$

$$\text{aa-AMP} + \text{tRNA} \rightarrow \text{aa-tRNA} + \text{AMP} \tag{2}$$

$$\text{aa-AMP} + \text{NDP/NTP} \rightarrow \text{Appp(p)N} + \text{Pi} \tag{3}$$

During UV irradiation cross-linking of s^4U somehow stimulates the production of AppppA and ApppA [41], but these dinucleotides are also synthesized by bacteria under other oxidative stress conditions and during heat shock, where s^4U state has not been evaluated [65–67,72]. Probably, an important source of the increase of these molecules concentration during heat shock is that the rate of dinucleotides synthesis by LysRS increases with temperature, while the rate of Lys-$tRNA^{Lys}$ synthesis decreases [64]. Also, the induction at higher temperatures of a specific LysRS isoenzyme (encoded by *lysU* gene) that is more stable and active for dinucleotides synthesis may explain the increase in AppppN during heat shock [64]. A possible explanation for the increase of these dinucleotides during oxidative stress could be the inactivation of their degradation enzyme, P1,P4-bis(5'-adenosyl) tetraphosphate pyrophosphohydrolase (ApaH), which *in vitro* is sensitive to cysteine oxidation [73]. An alternative is that oxidation of tRNA modified bases by itself activate the synthesis of the dinucleotides [65], similar to the case of UV irradiation where it has been shown that mutants that lack the s^4U modification cannot trigger the synthesis of AppppN [41].

Originally it was thought that these dinucleotides would function as "alarmones", that is, small molecules that are synthesized during a stress condition (in this case oxidative stress and heat shock) and trigger a fast response to it, similar to the role of (p)ppGpp during amino acid stringency (see below) [41,65–67,74]. Nevertheless, time course experiments showed that AppppA increase in concentration lags behind the synthesis of heat shock proteins during stress conditions, suggesting that it is not an alarmon in this case [72]. Also, over expression of *apaH*, the gene coding for AppppN hydrolase that eliminates these dinucleotides *in vivo* [73], decreased significantly the levels of ApppppA, but did not affected protein expression or cell survival under heat shock or H_2O_2 incubation [75]. Conversely, deletion of *apaH,* that increases the cellular level of ApppA, does affect cell survival and protein expression on several stress conditions including heat shock [76,77], UV irradiation [76], incubation with N-ethylmaleimide [76] as well during starvation [78]. It was shown that deletion of *apaH* somehow decreases expression of CAP-cAMP controlled genes which would decrease oxidative phosphorylation and limit further production of oxygen radicals. *apaH* mutant cells also showed prolonged synthesis of heat shock protein DnaK after heat shocked cells were returned to 30°C suggesting that the role of dinucleotides would be to modulate the long term response to stress conditions and not to trigger it [76]. The idea that the dinucleotides modulate the stress response is also supported by the fact that these molecules specifically bind several heat shock proteins including DnaK, GroEL[77] and ClpB [78]. Binding to DnaK inhibit its 5'-nucleotidase activity [79], but it is not known if it affects its ability to bind denatured proteins. Effects of ApppA binding to the other heat shock proteins has not been investigated, although it has been shown that increases in cellular level of Appp(p)N enhance the degradation of abnormal proteins synthesized during incubation with puromycin [78].

2.4. Role of aminoacyl-tRNA synthetases during oxidative stress

As discussed previously, LysRS and PheRS participate in the modulation of oxidative stress response through the synthesis of 5'-adenylyl dinucleotides. Many other aaRS that are oxidized during diverse oxidative stress conditions (Table 1) could also have relevant roles. Unfortunately, the effect of oxidation on most of these enzymes has not been characterized, although from some studied examples we know that oxidation can inactivate the enzyme [44] or increase its error rates [43]. In the case of human histidyl-tRNA synthetase (HisRS) apparently oxidation can even activate the enzyme [80], although we lack a thorough biochemical characterization of this oxidized enzyme.

In *E. coli,* cysteines from glutamyl-tRNA synthetase (GluRS) have been shown to be oxidized *in vivo* in cells lacking a periplasmic disulfuro isomerase (DsbA) [36]. The specific oxidized cysteines as well as the effect of their oxidation have not been characterized. Nevertheless, there are reports of the effect of *in vitro* oxidation on GluRS1 from *Acidithiobacillus ferrooxidans.* This enzyme has 4 cysteines one of which is near the active site, while the others form part of a Zn^{+2} binding domain. Oxidation of the Zn^{+2} binding cysteines inactivate the enzyme and release part of the Zn^{+2} [44]. Inactivation of the enzyme has been observed *in vivo* during excessive synthesis of tetrapyrroles like heme and it is supposed to regulate the synthesis of these molecules through modulation of the intracellular levels of its precursor, Glu-$tRNA^{Glu}$ [81]. Nevertheless, it is not known if the *in vivo* inactivation is due to oxidation and thus, if oxidative reactions are involved in this regulatory process.

The best characterized example of oxidation of a bacterial aaRS is the case of Threonyl-tRNA synthetase (ThrRS) which increases its error rates after *in vivo* or *in vitro* oxidation. ThrRS active site normally charges a small fraction of the $tRNA^{Thr}$ with serine (Ser). This Ser-

$tRNA^{Thr}$ could be potentially toxic to cells due to miss-incorporation of Ser in position of threonine (Thr) during protein translation. To prevent this mis-incorporation, ThrRS has an editing site that deacylate the mis-aminoacylated $tRNA^{Thr}$ [82]. Incubation of the enzyme with H_2O_2 inactivates its editing site, but does not affect the aminoacylation site. Thus, in these conditions the enzyme mis-acylation rate increases. Incubation of *E. coli* cells with H_2O_2 enhances the miss-incorporation of Ser in Thr positions confirming that oxidative stress do increase the error rates of the enzyme *in vivo* [43]. A similar example has been observed in eukaryotic cells where oxidative stress also increases the error rates of Methionyl-tRNA synthetase (MetRS), although instead of using the incorrect amino acid, the enzyme utilize non-cognate tRNAs [83]. Apparently, these mis-incorporation also happens in bacteria, where *E. coli* methionyl-tRNA synthetase (MetRS) is able to aminoacylate non-cognate tRNAs with Met, mainly $tRNA^{Arg}_{CCU}$ and $tRNA^{Thr}_{CGU}$ [84]. Unfortunately, there is no data with respect to the effect of oxidation on the mis-incorporation of Met in bacteria. It is expected that the mis-incorporation of Met could protect proteins from oxidative damage by oxidizing them before relevant targets [83]. This is in striking difference with the case of ThrRS, where it has been shown that mis-incorporation of Ser produces a lag in cell growth curves and a higher susceptibility to deletion of proteases [43].

2.5. Final remarks on the effects of oxidative stress on protein translation

While the increase of error rate by MetRS apparently protects the cell from further damage, a similar behavior by ThrRS decrease the fitness of cells to this stressful environment [43,83]. These contradictory effects of very similar phenomena are representative of what happens with all the components of the translation machinery and sometimes makes it difficult to determine whether an oxidized molecule is part of the control of protective mechanism or is itself a target of oxidation with deleterious consequences. At the same time, some of these effects can be deleterious or advantageous depending on the extent of the oxidative insult. An example of this is the role of s^4U modification on tRNA. The absence of this modification prevents the lag on replication after a short near UV irradiation and thus increases the cell fitness on these conditions. Nevertheless, when exposed for longer times to near UV, bacteria survive better when are able to produce the modification.

Cells cannot avoid using reactive molecules that are necessary for catalysis, but at the same time are easy targets of oxidative species. In many cases oxidation inactivates proteins that use this molecules and also destabilize their structure enhancing the formation of protein aggregates that are toxic to the cell. Thus, bacteria also need to have systems that protect them from further damage. An important step in cell protection is to prevent the synthesis of new proteins that could be targets of oxidation by inactivating several macromolecules involved in translation (Figure 1). At the same time, cells need to activate the translation of proteins that participate in the synthesis and usage of antioxidant molecules, as well as proteases and chaperons that prevent the formation of aggregates that are toxic to bacteria. After the stressful condition has passed, translation has to be re-initiated and proteases have to be inactivated in order to prevent the degradation of newly synthesized proteins that are folded correctly. The data presented in this chapter shows that the translation system has a central role in all this process. At one side is by itself a very important target of oxidation, which affects the rate of translation and its fidelity. This has profound effects on the cellular physiology some of which are protective and others deleterious. Also, as a target of oxidation, it appears that the translation system participate in the modulation of the response to stress. We still lack enough information in order to fully understand what is the specific effect of oxidation on each of the components of the system. At the same time, although we do have some hints, we do not understand how all these components interact between them during oxidative stress or how they coordinate with other oxidative stress response components.

3. Bacterial response to amino acid starvation

During evolution, living organisms have acquired various systems to survive under adverse environmental conditions. Upon nutrient starvation, bacteria slow down all processes related to cell growth and increase the functionality of processes that overcome nutrient deficit. This generalized process is known as the stringent response and occurs in cells designated as rel^{+}. The stringent response is triggered by the increase in the cellular levels of (p)ppGpp, also known as the "alarmone" or "magic spot" [85]. The level of these G nucleotide derivatives is regulated in *E. coli* by the activities of RelA and SpoT, two distinct but homologous enzymes. Under conditions of amino acid starvation, RelA senses uncharged tRNA stalled in the ribosome and synthesizes (p)ppGpp by pyrophosphorylation of GDP (or GTP) using ATP as the donor of pyrophosphate [86]. SpoT is a bifunctional enzyme that either synthesizes or degrades ppGpp [87] and its function is regulated in response to carbon, fatty acids or iron limitation. Catalytic activities in both enzymes are oriented to the amino termini regions that show conservation of the amino acid sequence. Conversely, carboxy termini are idiosyncratic since they are specific for each enzyme and their function is related to the signaling activities. Carboxy terminal domain (CTD) from RelA interacts with the ribosome probably sensing the uncharged tRNA [88]. SpoT contains in the CTD a region that interacts with the acyl carrier protein in the activation process [89]. RelA/SpoT related proteins have been found in all bacteria including the recently discovered "small alarmone synthetases" (SAS). These proteins seem to have complementary roles to RelA/SpoT. They are never alone but always in addition to RelA/SpoT in different combinations. RelA/SpoT homologous proteins have also been found in chloroplast probably with functions similar as in bacteria. Another ppGpp synthetase was found in chloroplasts of land plants that is sensitive to Ca^{++} [90]. Also in metazoan, another SpoT related protein, Mesh1, was recently identified [91]. The gene encoding Mesh1 compensates SpoT deficiencies in bacteria and *Drosophila* deficient in this protein show several impairments related to starvation. These findings widen the horizons of the functions of RelA/SpoT proteins to all kingdoms and also provide new relationships on signaling networks to control response to starvation.

3.1. Biosynthesis of (p)ppGpp in the ribosome

Aminoacyl-transfer RNAs (aa-tRNAs) are essential to cell physiology since they provide the amino acids to the ribosome for the translation of the genetic information encoded in the mRNA. Aa-tRNAs are synthetized by the aminacyl-tRNA synthetases and are delivered to the ribosome by the elongation factor EF-Tu (in bacteria) in the ternary complex aa-tRNA/EF-Tu/GTP. The ternary complex is positioned in the A site of the ribosome and as long as a correct pairing of the anticodon of tRNA with the codon in the mRNA is achieved, EF-Tu is released from the complex after hydrolysis of GTP. Upon formation of the peptide bond, the deacylated tRNA is released from the ribosome and is rapidly aminoacylated again by the corresponding aminoacyl-tRNA synthetase. Thus under normal growth conditions, the majority of tRNAs are aminoacylated and actively participating in protein synthesis. In contrast, under amino acid starvation, an important accumulation of deacylated tRNA takes place since the aminoacylation reaction is reduced. Under these conditions, an increasing number of A sites in the ribosome become empty or loaded with deacylated-tRNA and pausing of translation at these sites takes place [92-94]. Binding of deacylated-tRNA in the A site of the ribosome induce the formation of the RelA Activating Complex (RAC). RelA binds to RAC and catalyzes the transference of the β-γ pyrophosphate from ATP to either GTP or GDP for the formation of pppGpp or ppGpp respectively [93] (pppGpp is rapidly transformed to ppGpp, thus we will refer as ppGpp). Once ppGpp is formed, RelA is released from the ribosome but the deacylated tRNA might remain bound being released passively and independent from RelA. While deacylated-tRNA is still bound to the ribosome, it is unable to accommodate an incoming aa-tRNA/EF-Tu/GTP complex, thus it is stalled for protein synthesis. As long as RAC is active, new RelA molecules can bind and catalyze the formation of ppGpp [86]. As deacylated tRNA passively dissociates from the ribosome the stability of the interaction with the ribosome is a critical factor that influences the formation of ppGpp and thus the stringent response [95]. Recent data on the activation of RelA has shown that stalled ribosomes loaded with weakly bound deacylated-tRNAs require higher concentrations of enzyme than those loaded with tightly bound deacylated-tRNAs [96], suggesting that the recovery of cells from stringent response might be dependent on the type of starved amino acid.

3.2. Role of ppGpp in the transcription of stable RNAs and amino acids biosynthesis genes

The most well known effect of an increase in the concentration of ppGpp is the down regulation of the rRNA and tRNA transcription and thus of ribosomes and protein biosynthesis upon amino acid starvation. This is primarily an effect at the transcription level (reviewed in [97, 95]) and requires a direct interaction of the "alarmone" with the β and β′ subunits of the RNA polymerase affecting several activities, but mainly reducing transcription of rRNA genes. Biochemical, genetic and structural data indicate that ppGpp binds near the active site of RNA polymerase suggesting that the vicinity of this interaction might be involved in some of the observed effects [99-101]. There seems to be a reduced stability in the interaction between RNA polymerase and DNA in the open complex upon binding of ppGpp to the β and β′ subunits. Open complex at rRNA promoters is particularly unstable, thus this might be a requirement for the observed effect [102, 103]. However some stable open complexes are also affected by ppGpp suggesting that other mechanisms contribute to the effect of ppGpp in the activity of RNA polymerase at this level [104]. Other steps might be affected upon binding of ppGpp to RNA polymerase such as promoter clearance, open complex formation, pausing of transcription elongation and competition between ppGpp and other nucleotide substrates. These effects are not mutually exclusive and might take place at the same time.

Although the effect on stable RNAs is the major and the most well known effect on gene expression, a number of other functions related to cell growth are also affected upon ppGpp increase in the cell. Ribosomal proteins and elongation factors gene expression are negatively affected as well as fatty acids and cell wall biosynthesis. DNA biosynthesis is particularly sensitive to ppGpp and thus to amino acid starvation since in *E. coli* its progression stops soon after induction of ppGpp accumulation [105, 106]. ppGpp binds directly to DNA primase inhibiting initiation of DNA replication at both lagging and leading strands [107].

3.3. Role of DksA in the regulation by ppGpp

DksA is a protein that was discovered as a suppressor, when overexpressed, of the thermo sensitivity of *dnaK* mutants [108]. In addition it has many other functions, among these being the need of this protein and ppGpp to stimulate the accumulation of RpoS (the stationary phase and stress response σ factor) at the translational level [109]. A direct involvement of DksA potentiating the effect of ppGpp on the stringent response was discovered as one of its major functions [110, 111]. DskA is a structural homolog of the transcription elongator factors GreA and GreB [112]. These proteins bind directly to RNA polymerase particularly to the secondary channel of the enzyme inducing the cleavage of RNA in arrested enzymes rescuing them and restoring the polymerization activity. DskA seems to bind to RNA polymerase in a similar way, but without the induction of cleavage of RNA. Binding of DskA is believed to stabilize the interaction of RNA polymerase with ppGpp [112]. DksA can compensate the effect of a ppGpp^0 mutation (complete absence of ppGpp) reinforcing the notion that these two factors are synergistic both in positive and negative regulation. But DksA has also some other roles that are opposite to ppGpp, for instance in cellular adhesion, indicating that although compensatory, these two factors might have their own role in the stringent response [114].

Along with the pronounced inhibition of stable RNA transcription, positive effects on gene expression have also been observed upon increase of ppGpp levels. Two major ways to activate transcription have been proposed, direct and indirect activation. Direct activation implies the interaction of RNA polymerase with an efector such as ppGpp, DksA or both to activate transcription from a promoter. Transcription of several operons for the biosynthesis of amino acids, responding to the housekeeping σ^{70} factor, is activated by a direct mechanism. Promoters for the *hisG, thrABC* and *argI* are activated *in vitro* by a combination of ppGpp and DksA [112]. It is proposed that a step in the isomerization during the formation of the open complex is favored in the direct activation of these promoters.

Indirect activation of a specific promoter might be the result of the inhibition of other promoter, usually a strong one, that increases the availability of RNA polymerase to activate the target promoter [115, 116]. Evaluation of indirect activation of certain promoters comes mainly from *in vivo* studies. Activation of several σ factors other than σ^{70} also requires ppGpp. A competition mechanism that implies a reduced affinity of the core RNA polymerase for σ^{70} upon binding of ppGpp and/or DksA has been proposed, allowing to other σ subunits to bind to the core enzyme [117, 118]. It is speculated that RNA polymerase bound to strong promoters is released upon binding of ppGpp/DksA thus increasing the availability of the enzyme and also lowering the affinity to σ^{70} making the core enzyme available to the alternative σ factors.

In general speaking, ppGpp inhibits σ^{70} promoters of genes involved in cell proliferation and growth and activates promoters of genes involved in stress response and maintenance.

3.4. Targets for control of translation

The major effect of ppGpp in protein synthesis is certainly the biosynthesis of stable RNAs being inhibition of the transcription of rRNA and tRNA the targets for this effect. A marked reduction of the general translation of mRNAs as a result of the reduction of ribosomes as well as tRNAs is the major response against starvation of amino acids as well as other nutrients. In addition to this generalized response, other components of the translation machinery are also affected by the stringent response. Particularly translation factors that use guanine nucleotides are also target of ppGpp. As G proteins, these are the factors that have been the subject of analysis on the effect of (p)ppGpp at the translation level. G proteins are generally small proteins that bind GTP. The hydrolysis of this nucleotide, generally assisted by a G activating protein (GAP), to form GDP that remains bound to the protein, is required for the function to take place. The removal of GDP and its exchange for GTP is generally catalyzed by additional exchange proteins (GEP) that form part of the G proteins cycle [119].

Three proteins play important roles in the initiation step of translation, IF1, IF2 and IF3, being IF2 a G protein. IF3 binds to the ribosomal 30S subunit in the 70S ribosome releasing it from the 50S subunit to initiate a new cycle of elongation for the translation of an mRNA. IF1 assists IF3 in the releasing of the 30S subunit and also allows to the fMet-tRNA^{fMet} to be positioned in the correct P site to initiate translation. IF2 is a small G protein that in complex with GTP (IF2-GTP) binds the initiator fMet-tRNA^{fMet}. This ternary complex docks the fMet-tRNA^{fMet} in the small ribosome subunit. As the mRNA binds, IF3 helps to correctly position the complex such that the fMet-tRNA^{fMet} interacts by base pairing with the initiation codon in the mRNA. The mRNA is correctly positioned, assisted by the interaction of the Shine-Dalgarno sequence with the 16S rRNA, in the small 30S subunit. As the large 50S ribosomal subunit binds to the initiation complex, it participates as a GAP, thus GTP bound to the IF2 is hydrolyzed and released from the complex as IF2-GDP.

Elongation step of translation also requires in part the participation of the G-proteins EF-Tu and EF-G to take place. EF-Tu-GTP binds all aminoacyl-tRNAs with approximately the same affinity and delivers them to the A site of the ribosome in the elongation step of protein synthesis. Once a correct codon-anticodon interaction is detected by the ribosome, a conformational change in the ribosome takes place that induces the release of the EF-Tu factor along with the hydrolysis of GTP, thus the ribosome in this conformation acts as the GAP for the EF-Tu-GTP complex. EF-Ts is the GEP that assists EF-Tu-GDP to exchange GDP for GTP to initiate another elongation cycle.

EF-G is a G protein factor that complexed with GTP participates in the translocation of the nascent peptidyl-tRNA in the ribosome. Peptidyl transferase activity of the 23S RNA in the 50S subunit forms the peptide bond between the newly incorporated aminoacyl-tRNA in the A site delivered by EF-Tu and the existing peptidyl-tRNA already positioned in the P site from previous elongation cycles. The new peptidyl-tRNA with one extra amino acid is translocated from the A to the P site by EF-G-GTP. This process also implies the movement of the free tRNA positioned in the P site to the E site in the ribosome. EF-G itself seems to carry its own GEP.

RF3 releasing factor is also a G protein that participates in the termination of translation. Its function will not be discussed in this article.

As it is expected, these G proteins have been the subject of attention as potential targets for the action of ppGpp in the control of translation under the stringent response. GTP is at very high concentrations in the cell reaching more than 1 mM under normal growth conditions whereas GDP reaches very low concentrations. Upon amino acid starvation ppGpp can accumulate at the expenses of GTP that lowers its concentrations to nearly 50% [120-122]. Both nucleotides reach similar concentrations, thus depending on their affinities for the binding sites in proteins, they might compete. It is expected that G proteins can be severely affected under starvation since the levels of GTP are lowered, but also because ppGpp might interfere with its function. These proteins have been target of analysis since the early periods after discovery of the alarmone as the factor that influenced the stringent response. Initial studies indicated that pppGpp was able to substitute GTP in the reactions of EF2 and EF-Tu, but not in the function of EF-G [123]. Later studies revealed that EF-Tu as well as EF-G are inhibited by ppGpp, but this inhibition is dependent on the conditions of the reaction. EF-Tu is inhibited only if EF-Ts is not present. The inhibition can be fully reversed by the presence of aminoacyl-tRNA and EF-Ts [124].

As was mentioned before, IF2 is a G protein involved in the initiation of translation. This factor interacts in the initiation process with different ligands, ribosomal subunits, fMet-$tRNA^{fMet}$, GTP, GDP as well as ppGpp [125]. This protein participates in the entire initiation process and it has been shown by several methodologies that different conformational changes are necessary to each step. Because of the similar affinities of IF2 with GTP and GDP (dissociation constants between 10-100 μM), it is expected that under normal growth conditions (GTP 1 mM), IF2 binds the 30S subunit mostly with GTP bound. Hydrolysis of GTP, upon binding of the 50S subunit triggers the release of IF2-GDP from the initiation complex. Because this hydrolysis has not been proven as essential for this process, it led Milon et al. (2006) [126] to question the real role of this activity and asked about the reason for the evolutionary conservation of this process. The binding site of GTP and GDP in IF2, as well as in other G proteins involved in translation, is also the binding site for ppGpp. NMR data illustrates that ppGpp binds basically in the same site as GDP, although some differences might account for the different structure and function. To test the role of ppGpp on the IF2 function the authors measured the effect on different steps of the initiation process, i.e. binding of the fMet-tRNA^{fMet}, dipeptide formation, and the translation from the initiation codons on mRNAs containing AUG or AUU as initiator codon (the later being more dependent on IF2). All these steps in initiation of translation are severely affected upon ppGpp binding. From these studies the authors concluded that binding of ppGpp to IF2 might represent the signal to inhibit translation under conditions of metabolic shortage [126]. Thermodynamic analysis revealed that ppGpp binds to IF2 with higher affinity than to EF-G. Binding of fMet-tRNA^{fMet} to IF2 occurs with little variation in the presence of ppGpp compared to GTP while it is very sensitive to the nucleotides when complexed with the 30S subunit [127]. These results support the notion that initiation of translation is preferentially regulated by ppGpp under conditions of nutrient starvation.

3.5. Translation accuracy in the stringent response

Translational accuracy has been a topic of debate since the discovery of the stringent response. It is known that under amino acid starvation, *rel*$^+$ cells translation is at least 10 fold more accurate than in *rel*$^-$ although the rate of protein synthesis is the same in either type of cell. Different interpretations for this accuracy have been proposed, i.e. increased ribosome proof reading by ppGpp upon binding of either initiation or elongation factors, alterations of A site in the ribosome by binding of uncharged tRNA and different ribosome states controlled by the binding of ppGpp. It has also been proposed that there is no need for a special mechanism to maintain accuracy of translation since under amino acid starvation concentration of charged tRNA is not reduced as much as uncharged tRNA is increased [128]. At the same time, uncharged tRNA might bind to the A site in the ribosome competing for non-cognate tRNA thus reducing the chance to enter in the A site with the incorrect codon. Reduction in the activity of EF-Tu at the A site upon binding of ppGpp might reduce the chance to an error in translation. Measurements of aminoacylation levels for several tRNA revealed that *rel*$^-$ strains have at least five fold less aminoacyl-tRNAs than *rel*$^+$ strains suggesting that increased inaccuracy in these strains might be explained only by the charging level of tRNAs rather than other particular mechanisms [129]. These results imply that accuracy of translation is not affected under stringent response because there are either particular mechanisms that account for it or because there is a combination of effects (based on the real concentration of aminoacyl-tRNA, deacylated-tRNA bound to the A site of ribosome and the reduction in the translation rate by inhibition of IF2 and EF-Tu functions) that minimize the possibility that non-cognate aminoacyl-tRNAs enter the A site of the ribosome.

3.6. Overview of the effects of stringent response on translation

Upon amino acid starvation a generalized response, the stringent response, is achieved in bacterial cells. The major effector of this response is the marked increase of the cellular concentration of the nucleotide ppGpp, also known as the "magic spot" or the "alarmone".

This nucleotide is synthetized in the ribosome by the RelA protein upon activation by the presence in the A site of the ribosome of deacylated-tRNA. Two major effects on translation of the genetic information are observed. First, the dramatic reduction on the transcription of stable RNAs, i.e. rRNAs and tRNAs. The binding of ppGpp to the β and β′ of RNA polymerase triggers this effect by the destabilization of the open complex between RNA polymerase and strong promoters of stable RNAs. As consequence a marked reduction in the concentration of ribosomes and tRNAs slows down the translation of mRNAs. The second effect of the increase in the concentration of ppGpp on translation is an inhibition of translation itself by the effect on initiation and elongation steps. IF2, EF-Tu as well as EF-G are affected by the binding of ppGpp, but it seems likely that the initiation of translation through the inhibition of the IF2 function is the preferred target for the action of ppGpp to modulate the translation process. Accordingly, it has been proposed that IF2 might be a sensor to modulate translation depending on the nutritional status of the cell.

Author details

Assaf Katz
Department of Microbiology, Ohio State University, Ohio, USA

Omar Orellana
Program of Molecular and Cellular Biology, Institute of Biomedical Sciences, Faculty of Medicine, University of Chile, Santiago, Chile

Acknowledgement

This work was supported by grants from Fondecyt, Chile 1070437 and 111020 to OO and by University of Chile

4. References

[1] Sies H Role of reactive oxygen species in biological processes. Klinische Wochenschrift 199; 69: 965–8.
[2] Sies H Strategies of antioxidant defense. Eur. J. Biochem. 1993; 215: 213–9.
[3] Dukan S, Farewell A, Ballesteros M, Taddei F, Radman M, Nyström T Protein oxidation in response to increased transcriptional or translational errors. Proc. Natl. Acad. Sci. U.S.A. 2000; 97: 5746–9.
[4] McKenna SM, Davies KJ The inhibition of bacterial growth by hypochlorous acid. Possible role in the bactericidal activity of phagocytes. Biochem. J. 1988; 254: 685–92.
[5] Nathan C Nitric oxide as a secretory product of mammalian cells. Faseb J. 1992; 6: 3051–64.
[6] Rhee KY, Erdjument-Bromage H, Tempst P, Nathan CF S-nitroso proteome of Mycobacterium tuberculosis: Enzymes of intermediary metabolism and antioxidant defense. Proc. Natl. Acad. Sci. U.S.A. 2005; 102: 467–72.

[7] Barrette WC Jr, Hannum DM, Wheeler WD, Hurst JK General mechanism for the bacterial toxicity of hypochlorous acid: abolition of ATP production. Biochemistry 1989; 28: 9172–8.

[8] Nishiyama Y, Allakhverdiev SI, Murata N Protein synthesis is the primary target of reactive oxygen species in the photoinhibition of photosystem II. Physiol Plant 2011; 142: 35–46.

[9] Tamarit J, Cabiscol E, Ros J Identification of the major oxidatively damaged proteins in Escherichia coli cells exposed to oxidative stress. J. Biol. Chem. 1998; 273: 3027–32.

[10] Avery SV Molecular targets of oxidative stress. Biochem. J. 2011; 434: 201–10.

[11] Imlay JA Cellular defenses against superoxide and hydrogen peroxide. Annu. Rev. Biochem. 2008; 77: 755–76.

[12] Imlay JA Pathways of oxidative damage. Annu. Rev. Microbiol. 2003; 57: 395–418.

[13] Green J, Paget MS Bacterial redox sensors. Nat. Rev. Microbiol. 2004; 2: 954–66.

[14] Ghezzi P Oxidoreduction of protein thiols in redox regulation. Biochem. Soc. Trans. 2005; 33: 1378–81.

[15] Kiley PJ, Storz G Exploiting thiol modifications. PLoS Biol. 2004; 2: e400.

[16] Stadtman ER Oxidation of free amino acids and amino acid residues in proteins by radiolysis and by metal-catalyzed reactions. Annu. Rev. Biochem. 1993; 62: 797–821.

[17] Stadtman ER, Levine RL Free radical-mediated oxidation of free amino acids and amino acid residues in proteins. Amino Acids 2003; 25: 207–18.

[18] Yanagawa H, Ogawa Y, Ueno M Redox ribonucleosides. Isolation and characterization of 5-hydroxyuridine, 8-hydroxyguanosine, and 8-hydroxyadenosine from Torula yeast RNA. J. Biol. Chem. 1992; 267: 13320–6.

[19] Tanaka M, Han S, Song H, Küpfer PA, Leumann CJ, Sonntag WE An assay for RNA oxidation induced abasic sites using the Aldehyde Reactive Probe. Free Radic. Res. 2011; 45: 237–47.

[20] Nishiyama Y, Yamamoto H, Allakhverdiev SI, Inaba M, Yokota A, Murata N Oxidative stress inhibits the repair of photodamage to the photosynthetic machinery. Embo J. 2001; 20: 5587–94.

[21] Nishiyama Y, Allakhverdiev SI, Yamamoto H, Hayashi H, Murata N Singlet oxygen inhibits the repair of photosystem II by suppressing the translation elongation of the D1 protein in Synechocystis sp. PCC 6803. Biochemistry 2004; 43: 11321–30.

[22] Kojima K, Oshita M, Nanjo Y, Kasai K, Tozawa Y, Hayashi H, Nishiyama Y Oxidation of elongation factor G inhibits the synthesis of the D1 protein of photosystem II. Mol. Microbiol. 2007; 65: 936–47.

[23] Bosshard F, Bucheli M, Meur Y, Egli T The respiratory chain is the cell's Achilles' heel during UVA inactivation in Escherichia coli. Microbiology (Reading, Engl.) 2010; 156: 2006–15.

[24] Grant CM Regulation of translation by hydrogen peroxide. Antioxid. Redox Signal. 2011; 15: 191–203.

[25] Shan X, Chang Y, Lin CG Messenger RNA oxidation is an early event preceding cell death and causes reduced protein expression. Faseb J. 2007; 21: 2753–64.

[26] Honda K, Smith MA, Zhu X, Baus D, Merrick WC, Tartakoff AM, Hattier T, Harris PL, Siedlak SL, Fujioka H, Liu Q, Moreira PI, Miller FP, Nunomura A, Shimohama S, Perry G Ribosomal RNA in Alzheimer disease is oxidized by bound redox-active iron. J. Biol. Chem. 2005; 280: 20978–86.

[27] Shenton D, Smirnova JB, Selley JN, Carroll K, Hubbard SJ, Pavitt GD, Ashe MP, Grant CM Global translational responses to oxidative stress impact upon multiple levels of protein synthesis. J. Biol. Chem. 2006; 281: 29011–21.

[28] Amash HS, Brown OR, Padron VA Protection by selective amino acid solutions against doxorubicin induced growth inhibition of Escherichia coli. Gen. Pharmacol. 1995; 26: 983–7.

[29] Pericone CD, Park S, Imlay JA, Weiser JN Factors contributing to hydrogen peroxide resistance in Streptococcus pneumoniae include pyruvate oxidase (SpxB) and avoidance of the toxic effects of the fenton reaction. J. Bacteriol. 2003; 185: 6815–25.

[30] Brandes N, Rinck A, Leichert LI, Jakob U Nitrosative stress treatment of E. coli targets distinct set of thiol-containing proteins. Mol. Microbiol. 2007; 66: 901–14.

[31] Leichert LI, Gehrke F, Gudiseva HV, Blackwell T, Ilbert M, Walker AK, Strahler JR, Andrews PC, Jakob U Quantifying changes in the thiol redox proteome upon oxidative stress in vivo. Proc. Natl. Acad. Sci. U.S.A. 2008; 105: 8197–202.

[32] Mostertz J, Hecker M Patterns of protein carbonylation following oxidative stress in wild-type and sigB Bacillus subtilis cells. Mol. Genet. Genomics 2003; 269: 640–8.

[33] Dukan S, Nyström T Bacterial senescence: stasis results in increased and differential oxidation of cytoplasmic proteins leading to developmental induction of the heat shock regulon. Genes Dev. 1998; 12: 3431–41.

[34] Le H-T, Gautier V, Kthiri F, Malki A, Messaoudi N, Mihoub M, Landoulsi A, An YJ, Cha S-S, Richarme G. YajL, prokaryotic homolog of parkinsonism-associated protein DJ-1, functions as a covalent chaperone for thiol proteome. J. Biol. Chem. 2012; 287: 5861–70.

[35] Dukan S, Nyström T. Oxidative stress defense and deterioration of growth-arrested Escherichia coli cells. J. Biol. Chem. 1999; 274: 26027–32.

[36] Leichert LI, Jakob U. Protein thiol modifications visualized in vivo. PLoS Biol. 2004; 2: e333.

[37] Hu W, Tedesco S, McDonagh B, Bárcena JA, Keane C, Sheehan D. Selection of thiol- and disulfide-containing proteins of Escherichia coli on activated thiol-Sepharose. Anal. Biochem. 2010; 398: 245–53.

[38] Nawrot B, Sochacka E, Düchler M. tRNA structural and functional changes induced by oxidative stress. Cell. Mol. Life Sci. 2011; 68: 4023–32.

[39] Ramabhadran TV, Fossum T, Jagger J. In vivo induction of 4-thiouridine-cytidine adducts in tRNA of E. coli B/r by near-ultraviolet radiation. Photochem. Photobiol. 1976; 23: 315–21.

[40] Favre A, Hajnsdorf E, Thiam K, Caldeira de Araujo A. Mutagenesis and growth delay induced in Escherichia coli by near-ultraviolet radiations. Biochimie 1985; 67: 335–42.

[41] Kramer GF, Baker JC, Ames BN. Near-UV stress in Salmonella typhimurium: 4-thiouridine in tRNA, ppGpp, and ApppGpp as components of an adaptive response. J. Bacteriol. 1988; 170: 2344–51.
[42] Oppezzo OJ, Pizarro RA. Sublethal effects of ultraviolet A radiation on Enterobacter cloacae. J. Photochem. Photobiol. B, Biol. 2001; 62: 158–65.
[43] Ling J, Söll D. Severe oxidative stress induces protein mistranslation through impairment of an aminoacyl-tRNA synthetase editing site. Proc. Natl. Acad. Sci. U.S.A. 2010; 107: 4028–33.
[44] Katz A, Banerjee R, de Armas M, Ibba M, Orellana O. Redox status affects the catalytic activity of glutamyl-tRNA synthetase. Biochem. Biophys. Res. Commun. 2010; 398: 51–5.
[45] Kthiri F, Gautier V, Le H-T, Prère M-F, Fayet O, Malki A, Landoulsi A, Richarme G. Translational defects in a mutant deficient in YajL, the bacterial homolog of the parkinsonism-associated protein DJ-1. J. Bacteriol. 2010; 192: 6302–6.
[46] Kojima K, Motohashi K, Morota T, Oshita M, Hisabori T, Hayashi H, Nishiyama Y. Regulation of translation by the redox state of elongation factor G in the cyanobacterium Synechocystis sp. PCC 6803. J. Biol. Chem. 2009; 284: 18685–91.
[47] Kaiser II. Oxidation of 4-thiouridine in intact Escherichia coli tRNAs. Arch. Biochem. Biophys. 1977; 183: 421–31.
[48] Oppezzo OJ, Pizarro RA. Transient reduction in the tRNA 4-thiouridine content induced by ultraviolet A during post-irradiation growth in Enterobacter cloacae. J. Photochem. Photobiol. B, Biol. 2002; 66: 207–12.
[49] Oppezzo OJ, Pizarro RA. Inhibition of sulfur incorporation to transfer RNA by ultraviolet-A radiation in Escherichia coli. J. Photochem. Photobiol. B, Biol. 2003; 71: 69–75.
[50] Knox JP, Dodge AD. Singlet oxygen and plants. Phytochemistry 1985; 24: 889–96.
[51] Aro EM, Virgin I, Andersson B. Photoinhibition of Photosystem II. Inactivation, protein damage and turnover. Biochim. Biophys. Acta 1993; 1143: 113–34.
[52] Ejima K, Kawaharada T, Inoue S, Kojima K, Nishiyama Y. A change in the sensitivity of elongation factor G to oxidation protects photosystem II from photoinhibition in Synechocystis sp. PCC 6803. FEBS Lett. 2012; 586: 778–83.
[53] Lindahl M, Florencio FJ. Thioredoxin-linked processes in cyanobacteria are as numerous as in chloroplasts, but targets are different. Proc. Natl. Acad. Sci. U.S.A. 2003; 100: 16107–12.
[54] Auffinger P, Westhof E Appendix 5: Location and Distribution of Modified Nucleotides in tRNA. In: Grosjean H, Benne R, editors. Modification and Editing of tRNA, Washington, D.C.: ASM Press; 1998, pp. 569–76.
[55] Golovina AY, Sergiev PV, Golovin AV, Serebryakova MV, Demina I, Govorun VM, Dontsova OA. The yfiC gene of E. coli encodes an adenine-N6 methyltransferase that specifically modifies A37 of tRNA1Val(cmo5UAC). RNA 2009; 15: 1134–41.
[56] Murata M, Fujimoto H, Nishimura K, Charoensuk K, Nagamitsu H, Raina S, Kosaka T, Oshima T, Ogasawara N, Yamada M. Molecular strategy for survival at a critical high temperature in Eschierichia coli. PLoS ONE 2011; 6: e20063.

[57] Nyström T. Translational fidelity, protein oxidation, and senescence: lessons from bacteria. Ageing Res. Rev. 2002; 1: 693–703.

[58] Kramer GF, Ames BN. Oxidative mechanisms of toxicity of low-intensity near-UV light in Salmonella typhimurium. J. Bacteriol. 1987; 169: 2259–66.

[59] Favre A, Michelson AM, Yaniv M. Photochemistry of 4-thiouridine in Escherichia coli transfer RNA1Val. J. Mol. Biol. 1971; 58: 367–79.

[60] Yaniv M, Chestier A, Gros F, Favre A. Biological activity of irradiated tRNA Val containing a 4-thiouridine-cytosine dimer. J. Mol. Biol. 1971; 58: 381–8.

[61] Holler E, Baltzinger M, Favre A. Catalytic mechanism of phenylalanyl-tRNA synthetase of Escherichia coli K10. Different properties of native and photochemically cross-linked tRNAPhe can be explained in the light of tRNA conformer equilibria. Biochemistry 1981; 20: 1139–47.

[62] Thiam K, Favre A. Role of the stringent response in the expression and mechanism of near-ultraviolet induced growth delay. Eur. J. Biochem. 1984; 145: 137–42.

[63] Ramabhadran TV, Jagger J. Mechanism of growth delay induced in Escherichia coli by near ultraviolet radiation. Proc. Natl. Acad. Sci. U.S.A. 1976; 73: 59–63.

[64] Charlier J, Sanchez R. Lysyl-tRNA synthetase from Escherichia coli K12. Chromatographic heterogeneity and the lysU-gene product. Biochem. J. 1987; 248: 43–51.

[65] Bochner BR, Lee PC, Wilson SW, Cutler CW, Ames BN. ApppppA and related adenylylated nucleotides are synthesized as a consequence of oxidation stress. Cell 1984; 37: 225–32.

[66] Lee PC, Bochner BR, Ames BN. Diadenosine 5′,5‴-P1,P4-tetraphosphate and related adenylylated nucleotides in Salmonella typhimurium. J. Biol. Chem. 1983; 258: 6827–34.

[67] Lee PC, Bochner BR, Ames BN. AppppA, heat-shock stress, and cell oxidation. Proc. Natl. Acad. Sci. U.S.A. 1983; 80: 7496–500.

[68] Ibba M, Soll D. Aminoacyl-tRNA synthesis. Annu. Rev. Biochem. 2000; 69: 617–50.

[69] Goerlich O, Foeckler R, Holler E. Mechanism of synthesis of adenosine(5′)tetraphospho(5′)adenosine (AppppA) by aminoacyl-tRNA synthetases. Eur. J. Biochem. 1982; 126: 135–42.

[70] Plateau P, Blanquet S. Zinc-dependent synthesis of various dinucleoside 5′,5′ ″ ′-P1,P3-Tri- or 5′″,5′ ′ ′-P1,P4-tetraphosphates by Escherichia coli lysyl-tRNA synthetase. Biochemistry 1982; 21: 5273–9.

[71] Plateau P, Mayaux JF, Blanquet S. Zinc(II)-dependent synthesis of diadenosine 5′, 5‴ -P(1) ,P(4) -tetraphosphate by Escherichia coli and yeast phenylalanyl transfer ribonucleic acid synthetases. Biochemistry 1981; 20: 4654–62.

[72] VanBogelen RA, Kelley PM, Neidhardt FC. Differential induction of heat shock, SOS, and oxidation stress regulons and accumulation of nucleotides in Escherichia coli. J. Bacteriol. 1987; 169: 26–32.

[73] Guranowski A, Jakubowski H, Holler E. Catabolism of diadenosine 5′,5‴-P1,P4-tetraphosphate in procaryotes. Purification and properties of diadenosine 5′,5‴-P1,P4-tetraphosphate (symmetrical) pyrophosphohydrolase from Escherichia coli K12. J. Biol. Chem. 1983; 258: 14784–9.

[74] Varshavsky A. Diadenosine 5′, 5‴-P1, P4-tetraphosphate: a pleiotropically acting alarmone? Cell 1983; 34: 711–2.
[75] Plateau P, Fromant M, Blanquet S. Heat shock and hydrogen peroxide responses of Escherichia coli are not changed by dinucleoside tetraphosphate hydrolase overproduction. J. Bacteriol. 1987; 169: 3817–20.
[76] Farr SB, Arnosti DN, Chamberlin MJ, Ames BN. An apaH mutation causes AppppA to accumulate and affects motility and catabolite repression in Escherichia coli. Proc. Natl. Acad. Sci. U.S.A. 1989; 86: 5010–4.
[77] Johnstone DB, Farr SB. AppppA binds to several proteins in Escherichia coli, including the heat shock and oxidative stress proteins DnaK, GroEL, E89, C45 and C40. Embo J. 1991; 10: 3897–904.
[78] Fuge EK, Farr SB. AppppA-binding protein E89 is the Escherichia coli heat shock protein ClpB. J. Bacteriol. 1993; 175: 2321–6.
[79] Bochner BR, Zylicz M, Georgopoulos C. Escherichia coli DnaK protein possesses a 5′-nucleotidase activity that is inhibited by AppppA. J. Bacteriol. 1986; 168: 931–5.
[80] van Dooren SHJ, Raijmakers R, Pluk H, Lokate AMC, Koemans TS, Spanjers REC, Heck AJR, Boelens WC, van Venrooij WJ, Pruijn GJM. Oxidative stress-induced modifications of histidyl-tRNA synthetase affect its tRNA aminoacylation activity but not its immunoreactivity. Biochem. Cell Biol. 2011; 89: 545–53.
[81] Levicán G, Katz A, de Armas M, Núñez H, Orellana O. Regulation of a glutamyl-tRNA synthetase by the heme status. Proc. Natl. Acad. Sci. U.S.A. 2007; 104: 3135–40.
[82] Yadavalli SS, Ibba M. Quality control in aminoacyl-tRNA synthesis its role in translational fidelity. Adv Protein Chem Struct Biol 2012; 86: 1–43.
[83] Netzer N, Goodenbour JM, David A, Dittmar KA, Jones RB, Schneider JR, Boone D, Eves EM, Rosner MR, Gibbs JS, Embry A, Dolan B, Das S, Hickman HD, Berglund P, Bennink JR, Yewdell JW, Pan T. Innate immune and chemically triggered oxidative stress modifies translational fidelity. Nature 2009; 462: 522–6.
[84] Jones TE, Alexander RW, Pan T. Misacylation of specific nonmethionyl tRNAs by a bacterial methionyl-tRNA synthetase. Proc. Natl. Acad. Sci. U.S.A. 2011; 108: 6933–8.
[85] Potrykus, K., and Cashel, M. (p)ppGpp: still magical? Annu. Rev. Microbiol. 2008; 62: 35–51.
[86] Wendrich, T.M., Blaha, G., Wilson, D.N., Marahiel, M.A., and Nierhaus, K.H. Dissection of the mechanism for the stringent factor RelA. Mol. Cell 2002; 10: 779–788.
[87] Cashel M., Gentry D.R., Hernandez V.J., and Vinella D. The stringent response. In: Neidhardt F.C. (Ed. in chief), Escherichia coli and Salmonella: cellular and molecular biology, 2nd edition.ASM Press, Washington, DC, USA. 1996; pp. 1458–1496.
[88] Svitil AL, Cashel M, Zyskind JW. Guanosine tetraphosphate inhibits protein synthesis in vivo. A possible protective mechanism for starvation stress in Escherichia coli. J Biol Chem. 1993; 268(4): 2307-11.
[89] Battesti A., and Bouveret E. Acyl carrier protein / SpoT interaction, the switch linking SpoT dependent stress response to fatty acid metabolism. Molecular Microbiology 2006; 62: 1048–1063.

[90] Tozawa Y., Nozawa A., Kanno T., Narisawa T., Masuda S., Kasai K., Nanamiya H. Calcium-activated (p)ppGpp synthetase in chloroplasts of land plants. Journal of Biological Chemistry, 2007; 282: 35536–35545.

[91] Sun, D, Lee, G, Lee, J.H., Kim, H-Y, Rhee, H-W., Park, S-Y., Kim, K-J., Kim, Y., Bo Yeon Kim, B.Y., Jong-In Hong, J-I., Chankyu Park, Ch., Hyon E Choy, H.E., Kim, J.H., Young Ho Jeon, Y.H. and Chung, J. A metazoan ortholog of SpoT hydrolyzes ppGpp and functions in starvation responses Nat.Struct. Biol. 2010; 17: 1188-94.

[92] Dittmar, K. A., Sorensen, M. A., Elf, J., Ehrenberg, M., and Pan, T. Selective charging of tRNA isoacceptors induced by aminoacid starvation. EMBO Rep. 2005; 6: 151–157.

[93] Haseltine, W. A., and Block, R. Synthesis of guanosine tetra- and pentaphosphate requires the presence of a codon-specific,uncharged transfer ribonucleic acid in the acceptor site of ribosomes.Proc. Natl. Acad. Sci. U.S.A. 1973; 70: 1564–1568.

[94] Rojiani, M. V., Jakubowski, H., and Goldman, E. Effectof variation of charged and uncharged tRNA(Trp) levels on ppGppsynthesis in Escherichia coli. J. Bacteriol. 1989; 171, 6493–6502.

[95] Pedersen, F. S., Lund, E., and Kjeldgaard, N. O. Codon specific, tRNA dependent in vitro synthesis of ppGpp and pppGpp. Nat. New Biol. 1973; 243: 13–15.

[96] Payohe R. And Fahlman, R. Dependence of RelA-mediated (p)ppGpp formation on tRNA identity. Biochemistry 2011; 50: 3075-3083.

[97] Paul, B.J., Ross, W., Gaal, T. and Gourse, R.L. rRNA Transcription in Escherichia coli. Annu. Rev. Genet. 2004; 38, 749–770.

[98] Gralla, J.D. Escherichia coli ribosomal RNA transcription: regulatory roles for ppGpp, NTPs, architectural proteins and a polymerase-binding protein. Mol. Microbiol. 2005; 55, 973–977.

[99] Chatterji D, Fujita N, Ishihama A. The mediator for stringent control, ppGpp, binds to the betasubunit of Escherichia coli RNA polymerase. Genes Cells. 1998; 15: 279–87.

[100] Toulokhonov II, Shulgina I, Hernandez VJ. Binding of the transcription effector ppGpp to Escherichia coli RNA polymerase is allosteric, modular, and occurs near the N terminus of the β-subunit. J. Biol. Chem. 2001; 276: 1220–25.

[101] Artsimovitch I, Patlan V, Sekine S,VassylyevaMN,Hosaka T, et al. Structural basis for transcription regulation by alarmone ppGpp. Cell 2004; 117: 299–310.

[102] Kajitani, M. and Ishihama, A. Promoter selectivity of Escherichia coli RNA polymerase. Differential stringent control of the multiple promoters from ribosomal RNA and protein operons. J. Biol. Chem. 1984; 259: 1951–1957.

[103] Raghavan, A. and Chatterji, D. Guanosine tetraphosphateinduced dissociation of open complexes at the Escherichia coli ribosomal protein promoters rplJ and rpsA P1: nanosecond depolarization spectroscopic studies. Biophys. Chem. 1998; 75: 21–32.

[104] Potrykus, K, Wegrzyn G, Hernandez VJ. Multiple mechanisms of transcription inhibition by ppGpp at the lambda.p(R) promoter. J. Biol. Chem. 2002; 277: 43785–43791.

[105] Levine A., Vannier F., Dehbi M., Henckes G., Seror S.J. The stringent response blocks DNA replication outside the ori region in Bacillus subtilis and at the origin in Escherichia coli. Journal of Molecular Biology, 1991; 219: 605–613.

[106] Schreiber G., Ron E.Z., Glaser G. ppGpp-mediated regulation of DNA replication and cell division in Escherichia coli. Current Microbiology. 1995); 30: 27–32.

[107] Wang J.D., Sanders G.M., Grossman A.D. Nutritional control of elongation of DNA replication by (p)ppGpp. Cell. 2007; 128: 865–875.

[108] Kang PJ, Craig EA. Identification and characterization of a new Escherichia coli gene that is a dosagedependent suppressor of a dnaK deletion mutant. J. Bacteriol. 1990; 172: 2055–64.

[109] Brown L, Gentry D, Elliott T, Cashel M. DksA affects ppGpp induction of RpoS at a translational level. J. Bacteriol. 2002; 184: 4455–65.

[110] Paul BJ, Barker MM, Ross W, Schneider DA, Webb C, Foster, J.W. and Gourse, R.L DksA: a critical component of the transcription initiation machinery that potentiates the regulation of rRNA promoters by ppGpp and the initiating NTP. Cell 2004; 118: 311–22.

[111] Paul BJ, Berkmen MB, Gourse RL. DksA potentiates direct activation of amino acid promoters by ppGpp. Proc. Natl. Acad. Sci. USA 2005; 102: 7823–28.

[112] Perederina A, Svetlov V, Vassylyeva MN, Tahirov TH, Yokoyama S, Artsimovitch I, Vassylyev DG. Regulation through the secondary channel—structural framework for ppGpp-DksA synergism during transcription. Cell 2004; 118: 297–309.

[113] Vassylyeva MN, Svetlov V, Dearborn AD, Klyuyev S, Artsimovitch I, Vassylyev DG. The carboxy terminal coiled-coil of the RNAP β-subunit is the main binding site for Gre factors. EMBO Rep. 2007; 8: 1038–43.

[114] Magnusson LU, Gummesson B, Joksimovic P, Farewell A, Nystrom T. Identical, independent,and opposing roles of ppGpp and DksA in Escherichia coli. J. Bacteriol. 2007; 189: 5193–202 .

[115] Zhou, Y.N. and Jin, D.J. The rpoB mutants destabilizing initiation complexes at stringently controlled promoters behave like "stringent" RNA polymerases in Escherichia coli. Proc. Natl. Acad. Sci. U. S. A. 1998; 95: 2908–2913.

[116] Barker, M.M. Gaal T, Gourse RL. Mechanism of regulation of transcription initiation by ppGpp. II. Models for positive control based on properties of RNAP mutants and competition for RNAP. J. Mol. Biol. 2001; 305: 689–702.

[117] Laurie, A.D., Bernardo L.M., Sze, C.C., Skarfstad, E., Szalewska-Palasz, A., Nystrom, T. and Shingler, V. The role of the alarmone (p)ppGpp in sigma N competition for core RNA polymerase. J. Biol. Chem. 2003; 278: 1494–1503.

[118] Jishage, M., Kvint, K., Shingler, V. and Nystrom, T. Regulation of sigma factor competition by the alarmone ppGpp. Genes Dev. 2002; 16: 1260–1270.

[119] Caldon CE, Yoong P, March PE. Evolution of a molecular switch: universal bacterial GTPases regulate ribosome function. Mol Microbiol. 2001; 41(2): 289-97.

[120] Cashel M. The control of ribonucleic acid synthesis in Escherichia coli. IV. Relevance of unusual phosphorylated compounds from amino acid-starved stringent strains. J Biol Chem 1969; 244: 3133–3141.

[121] Cashel M, Gallant J. Two compounds implicated in the function of the RC gene of Escherichia coli. Nature 1969; 221: 838–841.

[122] Murray HD, Schneider DA, Gourse RL. Control of rRNA expression by small molecules is dynamic and nonredundant. Mol Cell 2003; 12: 125–134.

[123] Hamel E, Cashel M. Role of guanine nucleotides in protein synthesis. Elongation factor G and guanosine 5'-triphosphate,3'-diphosphate. Proc Natl Acad Sci U S A. 1973; 70: 3250-4.

[124] Rojas AM, Ehrenberg M, Andersson SG, Kurland CG. ppGpp inhibition of elongation factors Tu, G and Ts during polypeptide synthesis. Mol Gen Genet. 1984; 197: 36-45.

[125] Yoshida, M., Travers, A. & Clark, B. F. Inhibition of translation initiation complex formation by MS1. FEBS Lett. 1972; 23: 163–166.

[126] Milon, P., Tischenko, E., Tomsic, J., Caserta, E., Folkers, G., La Teana, A. Rodnina MV, Pon CL, Boelens R, Gualerzi CO. The nucleotide-binding site of bacterial translation initiation factor 2 (IF2) as a metabolic sensor. Proc. Natl Acad. Sci. USA. 2006; 103: 13962–13967.

[127] Mitkevich VA, Ermakov A, Kulikova AA, Tankov S, Shyp V, Soosaar A, Tenson T, Makarov AA, Ehrenberg M, Hauryliuk V. Thermodynamic Characterization of ppGpp Binding to EF-G or IF2 and of Initiator tRNA Binding to Free IF2 in the Presence of GDP, GTP, or ppGpp. J Mol Biol. 2010; 402: 838-46.

[128] Liljenstrom H. Maintenance of accuracy during amino acid starvation FEBS Lett. 1987; 223: 1-5.

[129] Sorensen, M.A. Charging levels of four tRNA species in Escherichia coli Rel(+) and Rel(-) strains during amino acid starvation: a simple model for the effect of ppGpp on translational accuracy. J Mol Biol. 2001; 307: 785-98.